Methods in Microbiology
Volume 28

Recent titles in the series

Volume 23 *Techniques for the Study of Mycorrhiza*
JR Norris, DJ Reed and AK Varma

Volume 24 *Techniques for the Study of Mycorrhiza*
JR Norris, DJ Reed and AK Varma

Volume 25 *Immunology of Infection*
SHE Kaufmann and D Kabelitz

Volume 26 *Yeast Gene Analysis*
AJP Brown and MF Tuite

Volume 27 *Bacterial Pathogenesis*
P Williams, J Ketley and G Salmond

Forthcoming titles in the series

Volume 29 *Genetic Methods for Diverse Prokaryotes*
MCM Smith and RE Sockett

Methods in Microbiology

Volume 28

Automation

Genomic and Functional Analyses

Edited by

Alister G. Craig

Molecular Parasitology Group
Institute for Molecular Medicine
John Radcliffe Hospital
Oxford
UK

and

Jörg D. Hoheisel

Functional Genome Analysis
Deutsches Krebsforschungszentrum
Heidelberg
Germany

ACADEMIC PRESS
San Diego London Boston
New York Sydney Tokyo Toronto

This book is printed on acid-free paper. ∞

Academic Press
24–28 Oval Road, London NW1 7DX, UK
http://www.hbuk.co.uk/ap/

Academic Press
a division of Harcourt Brace & Company
525 B Street, Suite 1900, San Diego, California 92101-4495, USA
http://www.apnet.com

A catalogue record for this book is available from the British Library

ISBN 0–12–521527–4 (Hardback)
ISBN 0–12–194860–9 (Comb bound)

Typeset by Phoenix Photosetting, Chatham, Kent
Printed in Great Britain by WBC Book Manufacturers Ltd, Bridgend, Mid Glamorgan
99 00 01 02 03 04 WB 9 8 7 6 5 4 3 2 1

Contents

Series Advisors

Gordon Dougan Department of Biochemistry, Wolfson Laboratories, Imperial College of Science, Technology and Medicine, London, UK

Graham J Boulnois Zeneca Pharmaceuticals, Mereside, Alderley Park, Macclesfield, Cheshire, UK

Jim Prosser Department of Molecular and Cell Biology, Marischal College, University of Aberdeen, Aberdeen, UK

Ian R Booth Department of Molecular and Cell Biology, Marischal College, University of Aberdeen, Aberdeen, UK

David A Hodgson Department of Biological Sciences, University of Warwick, Coventry, UK

David H Boxer Department of Biochemistry, Medical Sciences Institute, The University, Dundee, UK

Contributors

Gordana Aljinovic GATC-Gesellschaft für Analyse Technik und Consulting, Fritz-Arnold-Strasse 23, D-78467 Konstanz, Germany

JC Aude INRIA Rocquencourt, BP 105, 78153 Le-Chesnay Cedex, France

David R Bancroft Max-Planck-Institut für Molekulare Genetik, Ihnestraße 73, D-14195, Berlin-Dahlem, Germany

Helmut Blöcker GBF (Gesellschaft für Biotechnologische Forschung), Department of Genome Analysis, Mascheroder Weg 1, D-38124 Braunschweig, Germany

JJ Codani INRIA Rocquencourt, BP 105, 78153 Le-Chesnay Cedex, France

JP Comet INRIA Rocquencourt, BP 105, 78153 Le-Chesnay Cedex, France

Jean-Yves Coppee Centre de Génétique Moléculaire du Centre National de la Recherche Scientifique, Laboratoire Propre Associé à l'Université Pierre et Marie Curie, F-91198 Gif-sur-Yvette Cedex, France

Derrick WM Crook Department of Microbiology and Public Health Laboratory, John Radcliffe Hospital, Headington, Oxford OX3 9DU, UK

Amanda J Fife Department of Microbiology and Public Health Laboratory, John Radcliffe Hospital, Headington, Oxford OX3 9DU, UK

Dmitrij Frishman Munich Information Center for Protein Sequences/GSF, Am Klopferspitz 18a, 82152 Martinsried, Germany

Michael Y Galperin National Center for Biotechnology Information, National Library of Medicine, National Institutes of Health, Bldg 38A, Room 8N805, Bethesda, MD 20894, USA

E Glémet INRIA Rocquencourt, BP 105, 78153 Le-Chesnay Cedex, France

Alan N Hale Oxagen Ltd, Milton Park, Abingdon, Oxon, UK

Nicole C Hauser Functional Genome Analysis, Deutsches Krebsforschungszentrum, Im Neuenheimer Feld 506, D-69120 Heidelberg, Germany

A Hénaut Centre de Génétique Moléculaire du CNRS, Laboratoire Propre Associé à l'Université Pierre et Marie Curie, Avenue de la Terrasse, F-91198 Gif-sur-Yvette Cedex, France

Jörg D Hoheisel Functional Genome Analysis, Deutsches Krebsforschungszentrum, Im Neuenheimer Feld 560, D-69120 Heidelberg, Germany

Leroy Hood Department of Molecular Biotechnology, University of Washington School of Medicine, Box 357730, Seattle, WA 98195-7730, USA

Aneta Kaniak Centre de Génétique Moléculaire du Centre National de la Recherche Scientifique, Laboratoire Propre Associé à l'Université Pierre et Marie Curie, F-91198 Gif-sur-Yvette Cedex, France

Gerhard Kauer GBF (Gesellschaft für Biotechnologische Forschung), Department of Genome Analysis, Mascheroder Weg 1, D-38124 Braunschweig, Germany

Stephen Lasky Department of Molecular Biotechnology, University of Washington School of Medicine, Box 357730, Seattle, WA 98195-7730, USA

Hans Lehrach Max-Planck-Institut für Molekulare Genetik, Ihnestraße 73, D-14195, Berlin-Dahlem, Germany

Elmar Maier Max-Planck-Institut für Molekulare Genetik, Ihnestraße 73, D-14195, Berlin-Dahlem, Germany

Stefan Matysiak Functional Genome Analysis, Deutsches Krebsforschungszentrum, Im Neuenheimer Feld 506, D-69120 Heidelberg, Germany

Anthony J McCollum Imperial College of Science, Technology and Medicine, Mechanical Engineering Building, Exhibition Road, London SW7 2BX, UK

Gabriela Orlowska Institute of Microbiology, University of Wroclaw, 51-148 Wroclaw, Poland

Ulf Pettersson Department of Genetics and Pathology, Section of Medical Genetics, University of Uppsala, Biomedical Center, Box 589, S-75123 Uppsala, Sweden

Klaus-Jörg Rieger Centre de Génétique Moléculaire du Centre National de la Recherche Scientifique, Laboratoire Propre Associé à l'Université Pierre et Marie Curie, F-91198 Gif-sur-Yvette Cedex, France

JL Risler Centre de Génétique Moléculaire du CNRS, Laboratoire Propre Associé à l'Université Pierre et Marie Curie, Avenue de la Terrasse, F-91198 Gif-sur-Yvette Cedex, France

Lee Rowen Department of Molecular Biotechnology, University of Washington School of Medicine, Box 357730, Seattle, WA 98195-7730, USA

Marcel Scheideler Functional Genome Analysis, Deutsches Krebsforschungszentrum, Im Neuenheimer Feld 506, D-69120 Heidelberg, Germany

Piotr P Slonimski Centre de Génétique Moléculaire du Centre National de la Recherche Scientifique, Laboratoire Propre Associé à l'Université Pierre et Marie Curie, Avenue de la Terrasse, F-91198 Gif-sur-Yvette Cedex, France

Martin Vingron Theoretical Bioinformatics, Deutsches Krebsforschungszentrum, Im Neuenheimer Feld 280, D-69120 Heidelberg, Germany

A Wozniak INRIA Rocquencourt, BP 105, 78153 Le-Chesnay Cedex, France

Foreword

This book discusses a variety of modern techniques for deciphering biological information which employ powerful chemistries, instrumentation and analytic software in varying combinations. Over the past five years, biology has experienced a profound series of paradigm changes. For example, the Human Genome project has catalyzed the emergence of a new view of biology – namely, the idea that biology is an information science. This simple idea has profound implications for the practice of biology and medicine as we move toward the 21st century. There are three general types of biological information: the one-dimensional or digital information of DNA – the ultimate repository of life's information; the three-dimensional information of proteins, the molecular machines of life; and the four-dimensional (time-variant) information of complex biological systems and networks, such as the brain and the immune system. The digital information of DNA is diverse and represents a number of different chromosomal languages or distinct types of functional information, including those representing protein coding regions, regulatory elements, and the special features of the chromosome associated with its primary functions as an "information organelle". Sequencing DNA has two major objectives – to determine the prototype genome sequence for individual organisms and to understand the variation that occurs within each organism (polymorphisms) as well as how the genomes of organisms differ with respect to one another (comparative genomics). Proteins manifest their information by virtue of their shapes and chemical properties through their ability to interact with other molecules, often changing them or being changed by them. There are two major interesting problems with regard to proteins. The first, termed the protein folding problem, asks how one can determine the three-dimensional structure of a protein from its primary sequence of amino acid components. The second asks how the three-dimensional structure of individual proteins permits the execution of its function or functions. Biological systems and networks exhibit systems or emergent properties. For example, systems properties for the nervous system include memory, consciousness, and the ability to learn. Systems properties of the immune system include immunity, tolerance, apoptosis, and auto-immunity. The critical point is that systems properties emerge from the biological system functioning as a whole, that is the integrated interaction of its individual elements, and not from its individual components acting in isolation. For example, if one were to study a single neuron for 10–20 years and catalogue all of its activities, one would not learn one iota more about memory, consciousness, or the ability to learn, because these systems properties emerge from the operation of the network of neurons as a whole. Hence, biology must develop global tools to study all of the components of systems – a striking change from the last 30 years of biology where the focus of study was on the analysis of individual genes and individual proteins. It is still important to study individual genes and proteins: the point is that studying one protein or gene at a time will not lead to insights about systems properties. Two striking challenges for studying biological systems arise:

1. to develop high-throughput quantitative (global) tools for the analysis of biological information and.
2. to create models for these systems that accurately predict their systems properties

Deciphering biological information has two distinct meanings for each of the three types of information. On the one hand, one may decipher the human genome by determining the sequences of the 24 different human chromosomes. This is the objective of the Human Genome project. On the other hand, it is quite a different thing to discern the information that 3.7 billion years of evolution has inscribed in our chromosomes. This is the substrate of biology for the next 50–100 years. Likewise, with proteins, it is one thing to determine the three-dimensional structure of a protein and quite another to understand how that three-dimensional structure permits the protein to execute its particular functions. And so it is with biological systems – namely, it is one thing to define the elements and interconnections of the elements within the system, and it is quite another to understand how those elements and connections together give rise to the systems or emergent properties. A critical new component of biology will be the ability to collect and analyze systems information and the creation of models that will have the ability to predict how systems behave and give us deep insights into the nature of their emergent properties. This will require bringing applied mathematicians and computer scientists into close association with biologists, so that modeling based on detailed quantitative information for particular systems can be employed in the formulation of systems models.

◆◆◆◆◆◆ HIGH-THROUGHPUT ANALYTIC INSTRUMENTATION

The key to deciphering biological complexities, as noted earlier, lies in high-throughput analytic instrumentation and analysis falling into several major areas: genomics, proteomics, high-throughput phenotypic assays, high-throughput clinical assays, high-throughput multi-parameter cell sorting, combinatorial chemistry, and computational biology. This book focuses primarily on the techniques of genomics which include large-scale DNA sequencing, large-scale genotyping, DNA array analyses, and the attendant computational analyses required by large-scale genomic data. These techniques are moving beyond the execution of a single procedure to the creation of production lines which semi-automate and integrate a series of procedures. This process is nicely illustrated by large-scale DNA

* Numbers contained in parentheses throughout the Foreword refer to the relevant chapter numbers contained in this volume.

sequencing. This technique requires more than 50 different individual steps if one is to analyze in an ordered manner chromosomal DNA (7*). These steps include creation of a library of large insert fragments covering the entire genome, mapping procedures to create a minimum overlapping (tiling) path of these large insert clones across the individual chromosomes, random shearing of individual clone inserts and the construction of appropriate vector libraries, the plating out and picking of individual insert clones (2, 3), DNA preparation (4), DNA sequencing reactions, electrophoresis of the sequencing reactions, base calling, quality assessment of the bases, computational assembly of DNA fragments into the original insert sequence, finishing of these sequences to high-quality accuracy, annotation of the sequences and, finally, biological analysis of these sequences. Automation of large-scale DNA sequencing is going through several distinct steps:

1. the automation of the individual steps in the sequencing process
2. integration of as many individual steps as possible into a large-scale production line
3. the deployment of laboratory information management systems to control the production line and assess the quality of its performance.

In reality, the early and late steps of this process have not been fully automated nor integrated into a production line, but at many large genome centers the steps from colony picking through assembly and finishing have been semi-automated.

Large-scale genotyping focuses on the analysis of the DNA variation (e.g. single base substitutions, indels, etc.) that occurs within individual species (5, 6). We have now identified three generations of genetic markers: restriction fragment length polymorphisms (RFLPs), simple sequence repeats (microsatellites), and are moving toward the use of single-nucleotide polymorphisms (SNPs). Currently, most large-scale genotyping employs analysis of length variation in the highly polymorphic simple sequence repeats using gel sizing analyses often in the context of automated florescent DNA sequencers. About 20 000 simple sequence repeats are scattered across the human genetic map and more than 7000 across the mouse genetic map. The use of simple sequence repeats is somewhat limited by their high mutation rate. The single nucleotide polymorphisms have the advantage of lower mutation rates and ultimately the ability for very-high-throughput analyses via oligonucleotide arrays (see below), although they are less informative. In the next few years, the Human Genome project plans to generate 100 000 or more single-nucleotide polymorphisms scattered randomly across the genome, which may make it possible to identify genes predisposing to disease or normal physiological traits by direct association studies rather than analyzing families with the segregating trait which are difficult to collect. It will also facilitate the identification of weakly contributing modifier genes. The complete large-scale automation of simple sequence repeats or single-nucleotide polymorphism analyses represents a striking challenge not yet solved.

The DNA array technologies employ three distinct types of DNA attached to a solid matrix or surface – genomic DNA, cDNA and oligonucleotides (8). Each of these types of DNA arrays can be employed for different analyses; as more sequence data become available, however, the oligonucleotide arrays will become increasingly powerful as a tool for using the molecular complementarity of DNA to analyze a variety of features:

1. the expression patterns of all the genes in a genome with respect to biological, genetic or environmental perturbations;
2. single-nucleotide polymorphisms; and
3. resequencing DNA to identify interesting polymorphic variations.

Proteomics encompasses a second set of powerful tools that are just beginning to emerge. The idea is that complex mixtures of proteins can be separated (e.g. with two-dimensional gel electrophoresis or immunoprecipitation) and the resulting individual protein components analyzed very rapidly by mass spectrometry to determine protein identity, secondary modifications or even the ability of proteins to interact functionally with one another. The creation of high-throughput methodologies for the yeast two-hybrid system also gives one global capabilities for looking at protein interactions. Proteomics is moving rapidly toward the use of microfluidics and microelectronics to devise highly parallel and integrated technologies so that separated or chemically modified protein products can be analyzed by an appropriate analytic tool such as mass spectrometry. Many challenges remain in proteomics. Since with proteins there is no PCR to provide the capacity to analyze single molecules, a major question is how proteins that are expressed at very low levels can actually be visualized and characterized. A second issue has to do with solving the protein folding and structure–function problems as outlined above – the solution to these problems will in part be experimental and in part computational.

A variety of additional high throughput or global techniques will be invaluable in deciphering complex biological systems and networks. As global approaches are taken to destroy one at a time individual genes of yeast, nematode, drosophila and even mice, it will become even more imperative that we develop extremely high-throughput phenotypic assays for determining how these genetic perturbations have affected the informational pathways in these organisms (9). In a similar vein, it will be important to develop high throughput clinical assays (1). High-speed multiparameter cell sorting is going to be one of the real keys to understanding complexities of the nervous systems, the immune system, and developmental biology; it is only through characterization of individual types of cells and insights into how they change and execute their functions that we can come to understand how informational pathways operate. Combinatorial chemistry gives us the possibility to create enormous repertoires of molecules that can be used to perturb the biological information in cells and model organisms. The power of the diversity of the molecules made by combinatorial chemistry to decipher the intricacies of biology as well as to create the drug for medicine of the 21st century leads to another type of revolution in our ability to decipher biological information. Computational biology is about handling biological information and, of course, encompasses all of the above fields. The ability to analyze DNA (that is, to identify genes), carry out similarity analyses, identify repeat sequences, and so on, from entire genomes is strictly a computational task (10, 11). The ability to develop computational methods to solve the protein folding problem and predict accurately structure–function relationships is another important computational area. The ability to compare whole genomes of different organisms with each other and to infer how the informational pathways changed and, thus, to come to understand better their strategies for the execution of life is going to be a major opportunity denoted comparative genomics.

◆◆◆◆◆◆ FUTURE TOOLS

As I see it, there are four technologies that will provide exceptional opportunities for developing additional global tools for analyzing biological information. Microfluidics and microelectronics – more commonly abbreviated microfabrication – gives us the opportunity to create on silicon or glass chips the integration of multi-step processes as well as their parallelization and miniaturization. These techniques will to be the real key to creating next generation tools for genomics, proteomics and high-throughput phenotypic assays. Single-molecule analysis through scanning tip technologies, either at the DNA or at the protein level, offers enormous opportunities for the future. Indeed, it may be that large-scale DNA and sequence analysis in the future will be done at the single-molecule level. Single-molecule analysis will also allow us to look at the interaction of individual proteins with nucleic acids and other macromolecules. Nanotechnology affords an enormous opportunity for thinking about creating truly small molecular machines. The imagination can run wild as to what the nanomachines might do in terms of analyzing biological information. One area of critical tools in the future is going to be the analysis of informational pathways *in vitro* through various types of imaging procedures. To be able to look at the informational pathways as they operate in living creatures and come to understand both their systems connections and how these merge to give the emergent properties is going to be a very critical aspect of biology in the future.

Finally, the area of computational biology is only going to increase in importance as we develop better tools for capturing information, storing analyzing modeling, visualizing, and dispensing information. We must bring to biology virtually all tools of computer science and many of the tools of applied mathematics. The challenge of bridging the language barriers that separate biologists and mathematicians or computer scientists is an enormous one, but one that can be solved if biology is taught from this informational viewpoint.

In closing, this book gives a glimpse into where we are now with a variety of different computational and high-throughput analytic procedures. The future will be the development of new global strategies, more detailed integration of complex procedures, as well as their automatic control, and their miniaturization.

Leroy Hood, M.D., Ph.D.

Introduction

Microbiology is today advancing at an exceptional pace. This is chiefly a consequence of a series of technological breakthroughs. The major driving force is the ongoing human genome project. Microbiological diagnostics was for a long time dominated by rather traditional, mostly immunological, techniques and the potential for analysis of microbial genomes was not realised until many years after nucleic acid hybridisation was invented as a tool to identify nucleic acid sequences. In fact Gillespie and Spiegelman published their paper on solid phase nucleic acid hybridisation two decades before nucleic acid based methods were first used for microbial identification. While it was realised early that the genomes of viruses and bacteria harbour the information that makes every bacterial and viral species unique, the lack of defined probes and convenient detection methods hampered progress. One of the most important events in the history of microbiology was the invention of molecular cloning, allowing defined pieces of microbial genomes to be isolated and identified at the nucleotide level. Also the dramatic progress in nucleotide chemistry, which allowed synthetic oligonucleotide probes to be manufactured at a low cost, has been of key importance to the field. Another landmark in the history of molecular diagnostics was the invention of PCR. It is in fact today difficult to imagine how workers in the field could be sustained before the invention of this marvellous technology. PCR has certainly revolutionised nucleic-acid-based analysis by providing a simple method to generate highly specific targets. Moreover, since it allows detection of single molecules it offers an unsurpassed sensitivity. Yesterday's microbiological research was largely a manual business. The need to handle massive numbers of samples and clones in the Human Genome Project prompted the introduction of automation at many different levels. Progress has been extremely rapid and it is today a fact that many analytical steps in a modern biomedical research laboratory, ranging from colony picking to spotting of bacterial clones on filters, are carried out without the assistance of human hands.

The year 1977 was a landmark in the history of molecular biology since two entirely new methods for sequencing of nucleic acids were published that year, namely Maxam and Gilbert's method for sequencing by base-specific chemical degradation of end-labelled nucleic acids and Sanger's well-known dideoxy method. In the early days of molecular biology it was a major undertaking to sequence a dozen base pairs and the annual output of sequences before the mid-1970s was extremely limited and mostly the result of RNA sequencing by cumbersome methods. The new methodologies certainly changed the field and many laboratories started to produce sequences at speeds of thousands of nucleotides per year. Landmarks were the complete genome sequences of the phages ØX 174 and λ and the animal virus SV40. However, sequencing remained for many years a rather exclusive tool, mostly used to sequence limited regions where important genetic information was expected to be found.

The situation changed dramatically when the Sanger method became automated by machines that could read sequences from fluorescently labelled DNA products. Likewise the use of robotics for template preparation further improved speed and also precision.

The potential of high-throughput sequencing was realised by Venter and others who quickly embarked on the cDNA sequencing projects which have proven to be extremely useful for gene finding. Venter also realised that a well-managed sequencing facility would have the capacity to sequence complete bacterial genomes by a new and simple strategy. The year of 1995 will be remembered in the history of science as the year when the first sequence of a genome from a free living organism was reported. A new era in microbiology started and before the end of this century probably more than a hundred bacterial genomes will be completely sequenced. It is now fully appreciated that a wealth of information can be retrieved from genome sequences, allowing us to understand evolution and physiology in a new way. A problem which faces today's scientists is the need for massive parallel analyses. The introduction of solid phases on which thousands of clones or hundreds of thousands of oligonucleotides can be arrayed offers new unique opportunities to analyse vast numbers of sequences and to spot differences between them. The human geneticists are anxiously waiting for techniques which will allow the identification of thousands of polymorphisms in patient samples, thereby permitting the potential of association studies to be fully exploited.

Thanks to the technological progress mentioned above, an overwhelming amount of information is currently being gathered and a major challenge for the future is to develop tools to interpret this information. The need for new and more powerful informatic tools is continuously increasing and persons skilled in bioinformatics are for the moment a rare commodity. The genome projects will generate enormous amounts of descriptive information which will provide few clues to the function of the newly discovered genes. Much imagination will be needed to design methods that allow rapid analysis of gene function in a genome-wide perspective. Information already collected in the yeast and bacterial genome projects has demonstrated that much basic information about cell functions in complex organisms can be gained from simple unicellular organisms. Thus in the future microbial genomes will be studied not only in their own rights but also as models for the understanding of basic mechanisms in cellular function in general.

Modern microbiology requires a multidisciplinary mix of skills ranging from mechanical engineering to computer science. It is the hope that this volume will provide the reader with insights into some crucial areas of future microbial diagnostics.

Ulf Pettersson

1 Automation in Clinical Microbiology

Amanda J. Fife and Derrick W. M. Crook
Department of Microbiology and Public Health Laboratory, John Radcliffe Hospital, Headington, Oxford, UK

CONTENTS

I. INTRODUCTION

The process of automation in clinical microbiology is greatly influenced by its history and its position relative to other medical specialties. Clinical microbiology is a subspecialty of medicine which is laboratory based and is dedicated to the detection of infection by the analysis of clinical samples. It is distinct from but overlaps with infectious diseases which, in contrast, have a clinical base.

Clinical microbiology is one of a group of laboratory or pathology based specialties which includes clinical biochemistry, clinical immunology, clinical genetics, clinical haematology and histopathology which all historically arose from discrete areas of expertise. These specialties were highly differentiated as a result of their respective analytical methods and areas of human disease interest. In the past, many of the analytical techniques used by each laboratory discipline were manual. As a result, complex and unique methodological developments occurred in each specialty. This created the need for highly trained specialists, whose unique skills were a major impetus for the historical separation between the different disciplines of laboratory medicine.

Automation in microbiology has occurred largely through the development of new technologies which have been gradually assimilated in a piecemeal manner. To understand the opportunities for automation, it is

METHODS IN MICROBIOLOGY, VOLUME 28
0580-9517 $30.00

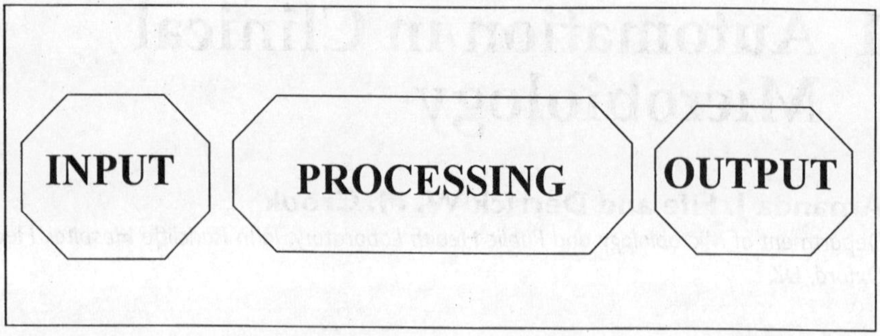

Figure 1. Three organisational elements to a microbiology laboratory.

helpful to analyse the full scope of what is considered to be a clinical microbiology laboratory.

The essential elements of a clinical microbiology laboratory consist of highly complex interrelated functions united in the common purpose of detecting infection. These can be greatly simplified and represented thus: first, the inputs to the laboratory such as the clinical specimens themselves and the laboratory consumables; second, the analytical processes; and, third, the outputs such as the reports to clinicians (Figure 1). Each of these components can in turn be further subdivided based on several discrete functional units (Figure 2) of which some are common to a number of laboratory subspecialties. Therefore, the traditional, largely methodological barriers which separate laboratory subspecialties begin to lose their relevance. The functional subdivisions of each of the components vary in the extent to which they are amenable to automation.

◆◆◆◆◆◆ II. STRUCTURE OF A CLINICAL MICROBIOLOGY LABORATORY

A. Inputs

The samples collected for analysis in a microbiology laboratory vary widely in terms of place of collection (GP surgery, hospital ward, operating theatre, etc.), time of collection, type of specimen and specimen "quality". Some types of specimen are of high quality, in that any positive culture from that specimen is likely to be diagnostically useful and clinically significant, whereas others are intrinsically of lower quality. Examples of the former include cerebrospinal fluid or pus collected during an operation. An example of the latter is the culture of expectorated sputum from hospitalised patients, the result of which usually reflects upper respiratory tract colonisation rather than identifying the aetiology of any pathological process in the lungs. A further important variable to be considered is the numbers of each specimen type submitted. As a result, it is impossible to predict on a day-to-day basis the numbers of each specimen type which will be received or the time at which they will

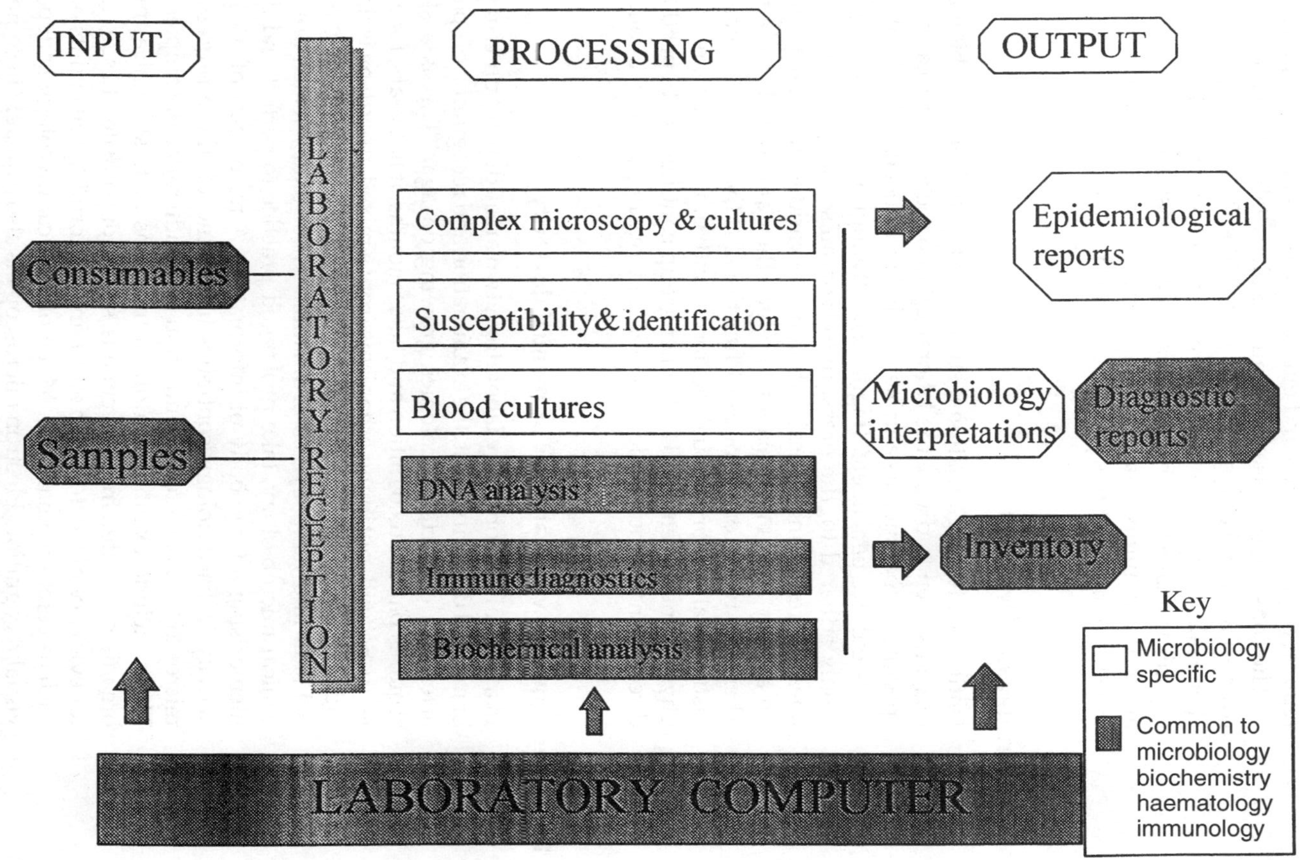

Figure 2. Organisation of a microbiology laboratory.

arrive, leading to uneven workflow in the laboratory. These variations are essentially common to all subspecialty areas but impose limitations on the extent to which this component of clinical microbiology is amenable to total automation. However, much of the essential data attached to a test request can be captured and entered on a laboratory computer.

The consumables used by a microbiology laboratory are extensive and overlap with other laboratory subspecialties. They include culture media, chemicals, immunological reagents and disposables. There are complex organisational issues in keeping a laboratory supplied with all its materials without unnecessary wastage. Apart from the automation of the inventory, much of this process remains manual.

B. The Analytical Process

Until now, the organisational structure of a clinical microbiology laboratory has been to divide the laboratory into sections or areas according to specimen type, such as a urine bench, faeces bench, blood culture bench, virus culture bench, etc. (Figure 2). Although this structure is still commonly adhered to, the pressure to reorganise along common functional lines is intensifying. Classification of the various functional units is helpful as they vary in the extent to which they are suitable for automation. There are two broad groups of functional activities: those which require visual analysis or manual dexterity and those which are suitable for physical measurement. The former group continues to depend largely on direct human input while the latter is increasingly being automated.

1. Processes requiring visual analysis or manual dexterity

In a clinical microbiology laboratory, two areas depend on visual analysis or manual dexterity. First, the examination and recognition of specific characteristics of bacterial colonies growing on agar. This is a skill which requires pattern recognition and takes months, if not years, for a person to learn. Second, purifying organisms from a mixed growth by isolating individual bacterial colonies (picking colonies) requires high degrees of manual skill and hand–eye co-ordination. These skills, which are unique to clinical microbiology, take prolonged practice to perfect and depend on memorising a large body of information. A major part of the laboratory activity in bacteriology continues to depend on these processes. Third, microscopy is used for examination of a wide range of samples and tests. These include examination of: Gram stains of fresh clinical material or organisms isolated from specimens; stools for parasites; tissue culture cells for evidence of a cytopathic effect and performing cell counts on samples such as cerebrospinal fluid. Much of medical mycology is dependent on visual recognition. Electron microscopy is also available in some laboratories to aid viral diagnosis. These activities share much in common with other specialties of pathology such as histopathology, cytology and haematology which also utilise microscopy extensively. The results from these processes are largely dependent on producing a descriptive written report which, again, increases the complexity over those processes which can produce a numerical result. Therefore, full laboratory automation for performing these analyses and producing a test result will depend on highly sophisticated image analysis, advanced artificial intelligence and robotics.

2. Processes requiring physical measurements

An increasing range of microbiology laboratory assays is dependent on the direct measurement of a physical characteristic. Many biochemical (including measurement of DNA or RNA) and immunological reactions can be measured colorimetrically, fluorometrically or photometrically. One example is the growth of bacteria in liquid media which can be measured by changes in density using a spectrophotometer. Also, simple images of bacterial growth on solid media can be detected and measured by commercially available video recorders and image analysers. These mechanically based measurements can be quantified and are suitable for automated systems. The results can be recorded in simple (usually numeric) codes directly by computer. There are also laboratory analytical processes which are shared by many of the laboratory subspecialties, such as clinical biochemistry, haematology, immunology and genetics, and which are ideal for automation. Equipment manufactured for these assays can be designed to undertake analysis of samples traditionally performed by separate laboratory specialties. Therefore, a single laboratory can be organised into units suitable for testing samples from multiple disciplines, based on the nature of the assay rather than the nature of what is being detected. For example, a laboratory organised along these lines may arrange a functional section or unit to undertake all immunodiagnostic assays. This unit would then perform all such assays for microbiology, immunology, haematology and clinical biochemistry. This particular laboratory arrangement lends itself to the scale of operation that produces significant economies favouring automation.

C. Outputs

There are three types of output from a microbiology laboratory. These are: first, diagnostic and screening test results; second, epidemiological reports which relate infection episodes between individual people, thereby detecting spread of an infectious disease in a population; and, third, reports providing measurements of laboratory performance (quality assurance). These outputs all depend on storage and analysis of data accumulated during the input and processing phases. Such data handling is ideally suited for computerisation.

1. Production of diagnostic and screening test results

Producing an analytical result is straightforward, but the interpretation of the result and determining the nature of the medical response are more complex. Also, the relative contributions of the laboratory and the clinician to this process vary between different types of test, different hospitals and different countries. The essential feature of this process is based on deducing the likelihood of a disease or infection in the person having the test. This is influenced by the false negative (sensitivity) and false positive (specificity) rate of the test and the prevalence of the condition in the population typical of the person being tested. Recording the test result with a simple interpretation is well within the capability of computer technology, but generating an automatic interpretation of all test results still remains beyond automated processes.

Control of infection in a hospital or the community depends on detecting episodes of infection that are linked. Examples of common infection control problems in which the microbiology laboratory plays an important detection and surveillance role are prevention of spread of epidemic strains of multiply resistant *Staphylococcus aureus* among vulnerable hospitalised patients and identifying the causative agents of community based food poisoning outbreaks. Searching either for specific patients or for isolates of specific organisms against computer databases containing laboratory test results is the most efficient method of abstracting this information. Similarly, analysis of laboratory performance depends on the ability of the laboratory information system to track, for example, the state of laboratory supplies, output of individual laboratory personnel and turn-round times for tests. This is easily recorded by and abstracted from a computer database.

◆◆◆◆◆◆ III. THE IMPACT OF AUTOMATION

A. Economic Issues

As in manufacturing industry, clinical laboratories are increasingly substituting labour intensive activities for automated processes. It is perceived that automation improves both the quality and the cost of the process. The main impact on quality is the elimination of test-to-test variation which arises from manual processes. Automated processes are usually capable of greater precision and reproducibility.

The perceived improvement in the cost of automation over manual processes has a major effect on the financial structure of the organisation. Labour intensive processes using highly skilled workers have high staff costs, whereas automated processes have high capital and consumable costs and may allow the employment of less skilled staff. Therefore, with progressive automation, the staff costs decrease and costs of consumables, maintenance and capital depreciation increase as a proportion of the operating budget. With investment in capital equipment the imperative is to maximise the return on investment. Operationally, the inevitable impact of this is to use equipment to full capacity which is the point at which the unit cost per test is likely to be lowest. Therefore, in pursuit of this ideal, equipment should be used continuously (i.e. 24 hours a day) and the volume of test throughput should approximate to the capacity. One obvious benefit to overall quality of capacity usage is the possibility of faster turn-round times for tests. Also, the larger the volume of consumables used, the greater the purchasing power of the organisation which then has the real potential of negotiating discounts which lower the unit cost of consumables. These factors lead to the inorexable pursuit of economies of scale, the consequence of which is the centralisation of laboratory activities.

Automation along these lines has a major effect on the organisation of a laboratory largely through altering the number, skill mix and working practices of staff. There is also a need for laboratory subspecialties to merge common processes to maximise the scale of the enterprise. This will enhance the economies that inevitably follow from a larger size of laboratory.

The major limitation to automation is mainly cost. The expense of developing or introducing new technologies may be greater than that of continuing with existing manual processes. Faced with the higher cost of automating a process, organisations are likely to choose the cheaper manual method. There are a number of areas where manual processes are likely to remain cheaper than automated ones. First, those processes which depend heavily on visual or fine manual dexterity, as the degree of technological refinement necessary to replace these processes would be prohibitively expensive. Second, tests for which the demand is low and the automated technology is both costly and unique to the test. In these circumstances, the unit cost per test is likely to be higher than that achievable by the equivalent manual process. Third, in countries where the cost of labour is low, manual systems may remain cheaper than what can be achieved by both economies of scale and automation.

B. Laboratory Computerisation

In a large laboratory a vast quantity of data is entered, recorded, stored, analysed and reported. This data is accumulated progressively as a sample is processed and passes through the input phase (patient details, specimen type, tests requested, etc.) and one or more of the functional units of the processing or output phases. The scope of handling this scale of information is beyond manual processes and the only means of achieving this is through computerisation. Improvements in the electronic transfer of information from machines to computers and between computers increases the scale of automation. The development and refinement of computer systems in hospitals and in general practices has also been important in facilitating this process. It allows the extension of automated processes closer to the point of taking the sample (computer generated requesting) and to reporting the result to the patient ("computer results reporting" to a ward or GP surgery). There are many vendors marketing and selling systems that provide integrated laboratory computer systems with excellent and improving performance which have the advantage that they can be applied to all the subspecialties.

C. The Input Phase

Automation of this phase is mainly achieved by computerisation of specimen requesting or ordering laboratory supplies. These activities are common to all the subspecialties and can be merged.

D. The Processing Phase

This can be organised into a number of functional units, each of which may be automated. Functions shared with other subspecialties can be merged (these account for approximately 30–50% of the throughput of a standard hospital based laboratory).

1. Processes common to different subspecialties

(a) Immunodiagnostics

Enzyme-linked immunosorbent assay (ELISA) technology is applied to a wide range of assays traditionally performed by clinical microbiology, clinical immunology, haematology and clinical biochemistry. The method can be applied to any molecule capable of participating in a specific antibody–antigen reaction. The development of monoclonal antibody technology has improved the sensitivity and specificity of the technique. Large automated ELISA analysers are commercially available and are capable of performing tests for most of these subspecialty areas on a single machine. One example of the application of ELISA technology to clinical microbiology has been in the detection of infection with *Chlamydia*. *Chlamydia trachomatis* infects and colonises the human genital tract and contributes to the pathogenesis of pelvic inflammatory disease in women which results in pain and infertility. It can also cause serious eye and lung infections in babies born to women with active infection. Screening for the presence of this organism in populations likely to have been exposed is important as treatment is available. Culture of these obligate intracellular organisms requires the use of tissue culture, facilities for which not all diagnostic laboratories possess. Culture remains the reference method to which other methods are compared. The introduction of an ELISA, such as the Syva MicroTrak System, to detect chlamydial antigen in genital specimens from patients with symptoms or those attending genitourinary medicine clinics has made automated screening possible. As with any diagnostic test, the cut-off value for a positive result has to be determined in order to give acceptable sensitivity and specificity and the interpretation of the result has to be made in the light of the pre-test probability of the patient having the condition. This means that the test may be less reliable if a low prevalence population is screened, thus routine screening is not offered to all pregnant women. Repeat or further confirmatory testing is required in the case of the equivocal or unexpected result. A DNA probe based assay which detects chlamydial rRNA (Gen-Probe PACE 2) appears to give similar results to ELISA and is also suitable for automated screening. Improvements in the sensitivity and specificity of screening may occur as a result of the introduction of DNA based technology such as the ligase chain reaction (LCR) for detection of *C. trachomatis* (see (b) below).

(b) DNA based assays

Detection of specific DNA sequences is central to most of these tests. The most powerful of these techniques are those based on the amplification of specific DNA sequences and include the polymerase chain reaction (PCR) and LCR. The latter has been successfully adapted to the detection of *C. trachomatis* in clinical samples and offers improved sensitivity over ELISA and DNA probe based methods. The automation of this technology is still underway, but is rapidly advancing. Once it is refined, it is likely that large automated analysers based on technologies described elsewhere in this book will be capable of undertaking assays for many subspecialties including clinical microbiology. Developments in this area are likely to make a major contribution to the diagnosis of infections with slow growing or unculturable organisms where the current methods give either indirect or retrospective evidence of infection (for instance, the detection of an antibody reaction to the infective agent). DNA based assays have the advantages of being very sensitive and highly specific when performed under the correct conditions.

(c) Biochemical assays

An increasing range of automated tests is being developed which are replacing manual processes traditionally performed in a clinical microbiology laboratory. Measurement of antibiotic concentrations in body fluids is well suited to automated equipment commonly used in clinical biochemistry. Recently, automated biochemically based indicator strips or "dip-stix" assays of urine are replacing the need for most urine microscopy. Previously, microscopy of urine was a highly labour intensive and skilled process which took up a considerable part of the time required actually to process the specimen. As a large clinical laboratory may receive upwards of five hundred specimens a day, the use of strips which indicate the presence of leucocyte esterase and/or nitrites (from bacterial reduction of nitrates), correlating with the leucocyte count and the presence of bacteria respectively, represents a considerable saving in staff time, especially as the reading of the strips is automated. It is likely that simple biochemical assays will increasingly replace what were previously manual assays.

2. Processes specific to microbiology

These account for 50–60% of a hospital laboratory throughput. The following processes are considered: automation of blood cultures; automated antimicrobial susceptibility testing and identification; tasks requiring a high level of dexterity or pattern recognition.

(a) Automated blood culture machines

The culture of organisms from the blood of sick or febrile patients is one of the most important roles of the clinical microbiology laboratory. The blood of healthy individuals should be bacteriologically sterile, therefore any organism cultured could potentially signify bacteraemia. However, there is also a significant contamination rate, almost always as a result of sampling technique but occasionally as a result of post-sampling processing. Blood obtained by aseptic technique is inoculated into paired blood culture bottles (one aerobic, one anaerobic) containing broth designed to support the growth of a wide range of organisms. The established manual method of blood culture processing (still in use in many countries) is labour intensive and depends on the visual examination of the bottles for macroscopic evidence of bacterial growth. This manual approach depends on the preparation of Gram stains from "suspect" bottles. All bottles are subcultured on to solid media at 2 days and again at the end of incubation, after 5–7 days. This process is subject to considerable operator variation and there is an ever present possibility of introducing contaminating bacteria during repeated manipulation. Automated methods for detecting growth in blood culture bottles have been developed which are both sensitive and standardised, as well as being non-invasive. They are based on the physical detection of the metabolic products of bacterial growth, usually carbon dioxide. The earliest automated systems to be widely used were the Bactec Radiometric systems which utilised media with radiolabelled carbon-containing substrates. Bacterial growth generated radiolabelled carbon dioxide which could be detected by monitoring the composition of the bottle headspace gas, usually twice daily. Bottles exceeding the threshold value could then be Gram stained and subcultured. Refinement of the early systems has resulted in the development of machines which can provide continuous non-invasive monitoring using a variety of detection methods which do not involve radioactivity. The most recently commercially available Bactec systems, such as the Bactec 9240 developed by Becton Dickinson (one of a number of vendors supplying automated equipment of similar quality) (Figure 3), have sensors in the base of each bottle which respond to rising carbon dioxide concentrations in the liquid medium by producing changes in fluorescence which can be detected fluorometrically. Other commercially available continuous monitoring systems utilise colorimetric detection (BacTAlert, Organon Tecnika) or headspace gas pressure changes (ESP, Difco). The bottles are monitored every 10 minutes. The computer algorithms of this later generation of automated machines are designed to detect both absolute levels and changes in levels over time of the parameter which is being monitored. The setting of the detection thresholds for positive blood cultures has to be carefully balanced when commissioning the system. Obviously, it is most important that no genuine positive cultures go undetected. However, if the threshold levels are too low, then the false positive rate may be too high (large numbers of white blood cells in a sample are a common cause of false positive signals), leading to an unacceptably high workload. As with all automated systems, a balance

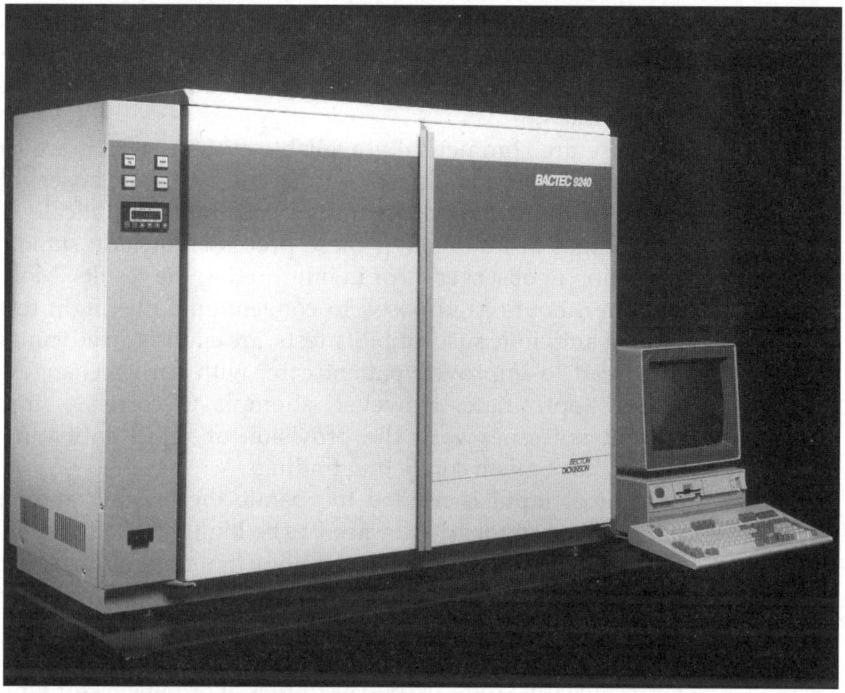

Figure 3. The Bactec 9240 automated blood culture machine (Photograph courtesy of Becton Dickinson).

between sensitivity and specificity has to be struck. Although all incubation and monitoring is automated with modern systems, bottles flagged as positive still require manual and visual processing to make and examine the Gram film and to subculture the bottles appropriately on to solid media.

(b) Automation of identification and susceptibility testing

When an organism which is likely to be clinically significant is isolated from blood or other clinical material, it is necessary to provide data on the identity and antimicrobial susceptibility pattern of the organism for two main reasons. First, this information allows antimicrobial treatment to be optimised for individual patients and the identification of the organism may even give a clue to the aetiology of a patient's condition in specific cases. An example of this is the finding of *Streptococcus bovis* in the blood cultures of a patient, as this organism is associated not only with infective endocarditis, but also with the presence of bowel malignancy. Second, it is important to know the identity and common antibiotic susceptibility patterns of organisms circulating in the hospital setting for surveillance and infection control purposes. An example of this would be the unexpected appearance of a highly resistant strain of *Klebsiella* in good quality specimens from patients in an intensive care unit where this organism was not previously endemic. This would alert infection control staff to the need for isolation procedures and may also necessitate a change in the empiric antibiotic regimes until such a time as the outbreak is controlled.

There are commercially available systems which are capable of performing simple identification tests and antibiotic susceptibility tests on many of the common organisms encountered clinically. The potential advantages of automating these processes include standardisation and reduction in observer error in interpreting the results. More rapid results (within 6 hours as opposed to conventional overnight testing), particularly of antibiotic susceptibility tests, are cited as an advantage in that they may lead to improved patient care, with earlier changes in antibiotics where appropriate. However, attempts to correlate improvements in patient outcomes with the provision of rapid antibiotic susceptibility results have given conflicting findings.

Manual input is needed to operate the existing automated systems, although operators do not need to be highly skilled. The main limitation to the extensive use of this technology in many countries is the expense of the capital outlay and consumables. Also, existing technology in this field, despite continuous refinement, is still limited in the range of organisms which can be reliably tested. Fastidious and non-fermenting aerobic Gram-negative rods are two examples of organisms for which it has so far been necessary to maintain manual systems. The interpretation of the results produced by automated machines remains a skilled manual process, although some systems are capable of limited interpretation, based on rules. This is useful for ensuring that reports with unusual or unacceptable antibiotic susceptibility profiles are intercepted. Manual based processes remain cheaper in many countries where the cost of labour is low.

The identification of organisms by automated systems is based on detection of biochemical reactions or substrate utilisation in liquid media and is therefore not dissimilar in principle from commercial identification kits such as the API system (Biomérieux) which have been available for manual use for many years. The reagents are provided in multi-well trays to which a standard inoculum of the bacterium is added. The susceptibility of the organism to a battery of preselected antibiotics is similarly determined by measuring liquid phase growth in wells. The wells are pre-inoculated with predetermined quantities of each antibiotic which when inoculated with a set volume of bacterially seeded broth generates a "dilution series" of each antibiotic, enabling the minimum inhibitory concentration or MIC of the organism to antibiotics to be determined. The MIC is the concentration of antibiotic required to inhibit the growth of an organism and allows predictions to be made about the likely clinical response of a patient with an infection at a particular site to a given antibiotic. This is based on what is known about achievable serum levels and the distribution of the antibiotic in different anatomical sites of the body (such as lungs, kidney, etc.). Measurement of biochemical reactions or bacterial growth is based on the detection of a colour change or optical density of a liquid medium, respectively, by spectrophotometry.

Identification of organisms or determination of susceptibility can also be performed by video-recording and image analysis, with similar results. The identification system using this technology is based on multi-point inoculation of solid media containing specific biochemical indicators or a dilution series of an antibiotic. The agar plates can be read and the organism identified based on the pattern of its biochemical reactions. Its antibiotic susceptibility pattern can be determined on the basis of inhibition of growth (colony formation) by a known concentration of antibiotic in the medium. This represents the MIC of the organism to the antibiotic. One of the most commonly used manual methods of antibiotic susceptibility testing is to place antibiotic impregnated discs on solid media seeded with test and control organisms, using inocula of standard density. Susceptibility of test organisms can be determined by measuring the size of the inhibition zone around the disc and comparing it with the zone of the control organism. Automation can be introduced to control both the inoculum size and zone size measurement, thus increasing the standardisation of the procedure. Zone size analysis can be performed using video-recording and image analysis to give accurate results.

(c) Processes requiring high degrees of visual or manual skill

As discussed earlier, many culture based analyses of samples are still dependent on highly skilled manual processes. Good examples of these are the interpretation of cultures from body sites with a mixed indigenous microbial flora such as sputum or faeces, where the skill lies in recognising and purifying the pathogen from a mixed background, for instance the ability to recognise small numbers of *Salmonella* in a mixed stool culture. Another example is the recognition of common intestinal parasites by microscopy, as there is no satisfactory rapid automated alternative. Advances in automation in this area have been limited, but entering data directly on to the computer has reduced the need for paperwork and has also reduced the chance of clerical or transcription errors.

E. The Output Phase

Test results with simple interpretations can be readily reported by computer. These reports can be relayed electronically to their destination. Analysis of data relevant for infection control, management of the laboratory and maintaining the stock of the laboratory can also be readily produced by computer. These processes are common to all the laboratory subspecialties.

The detailed and complex interpretations of test results remain largely dependent on manual and specialist medical input. These interpretations can be directly recorded on a computer. It is in this area that specialist medical expertise is focused in a large laboratory where many analytical functions which were traditionally separated are now merged.

◆◆◆◆◆◆ IV. THE FUTURE OF AUTOMATION

The major impact of automation is to remove the subspecialty barriers and thus move towards larger laboratories operating continuously near to full capacity. These laboratories are likely to be subdivided along functional lines rather than by subspecialty. Those elements of clinical microbiology which are based on a high degree of skill, visual and hand–eye co-ordination will not be automated in the foreseeable future, unless image analysis, artificial intelligence and robotics improve in performance and substantially reduce in cost. These manual processes will remain a relatively small functional unit of their own. The ideal of a completely automated clinical microbiology laboratory has not been achieved and may never be feasible.

The major alternative to automation, based on centralisation of laboratory activity into large units, is the development and expansion of near patient testing. The advancement in rapid tests, based on highly reliable micro-technology suitable for the bedside or doctor's office, raises the spectre of these technologies replacing large centralised laboratories, particularly as such tests inevitably give faster results than those performed by the laboratory. Some of these tests can be automated but many remain dependent on manual interpretation. The main factors which will determine the evolution of near patient testing will be the demonstration of sensitivity and specificity equivalent to the laboratory test, and the unit cost of the "bedside" test compared with the laboratory test. It must be remembered that the quality of the near patient test will depend not only on the intrinsic robustness of the test itself but also on storing and using the kit in accordance with the manufacturer's specifications, which is easier to control in the laboratory setting than in wards and physicians' offices.

◆◆◆◆◆◆ V. SUMMARY

It is apparent that the drive to automation in clinical microbiology will continue with an increasing proportion of the work of the laboratory being replaced or augmented by automated systems. Advances, particularly in DNA amplification based methods and immunodiagnostics, will increase the ability of the laboratory to detect infections with organisms for which there has previously been no satisfactory diagnostic test. This will ultimately benefit the patient in terms of earlier specific diagnosis and, therefore, earlier specific treatment.

Replacement of processes in clinical microbiology which require high levels of interpretative, visual and manual skill is likely to be much slower and some may remain irreplaceable. Advances in existing technologies such as PCR may replace some processes by becoming recognised diagnostic methods for some of the infective organisms currently detected by conventional means.

Should robotic technology and artificial intelligence ever reach a state where the total automation of clinical microbiology becomes feasible, the question of whether there would be a continuing need for human involvement is an interesting one and would depend on two factors: first, a continuing demand for human input from the users of the service; and, second, whether computers with artificial intelligence will ever have the extra creative dimension required to make judgements and management decisions in unique situations.

Further Reading

Clarke, L. M., Sierra, M. F., Daidone, B. J. *et al.* (1993). Comparison of the Syva MicroTrak enzyme immunoassay and Gen-Probe PACE 2 with cell culture for the diagnosis of cervical *Chlamydia trachomatis* infection in a high prevalence female population. *J. Clin. Microbiol.* **31**(4), 968–971.

Doern, G. V., Vautour, R., Gaudet, M. *et al.* (1994). Clinical impact of rapid *in vitro* testing and bacterial identification. *J. Clin. Microbiol.* **32**(7), 1757–1762.

Jorgensen, J. H. (1993). Selection criteria for an antimicrobial susceptibility testing system. *J. Clin. Microbiol.* **31**(11), 2841–2844.

Stanek, J. L. (1995). Impact of technological developments and organisational strategies on clinical laboratory cost reduction. *Diag. Microbiol. Infect. Dis.* **23**, 61–73.

Wilson, M. L., Weinstein, M. P. and Reller, L. B. (1994). Automated blood culture systems. *Clin. Lab. Med.* **14**(1), 149–169.

2 Vision Systems for Automated Colony and Plaque Picking

Anthony J. McCollum
Imperial College of Science, Technology and Medicine, Exhibition Road, London, UK

CONTENTS

LIST OF SYMBOLS

+	addition, logical OR
−	subtraction
×	multiplication
÷, /	division
●	logical AND
m	Boolean variable
$\bar{m}$	logical NOT
E, E_1, E_2	luminous flux
F	illumination or illuminance
π	pi
θ	angle theta
r, a	radial distance
A	area
$\int_a^r r\,dx$	integral with respect to dx
$\left[x\right]_a^r$	definite integral
L, H, l, v	distance
f	focal length
$p(x, y)$, $f(x, y)$	function p, f of variables x, y
W	Maximum grey level (white)
≤	less than or equal to
≥	greater than or equal to

METHODS IN MICROBIOLOGY, VOLUME 28
0580-9517 $30.00

I, J	array dimensions
$p_{i,j}$	pixel p at point i, j
[0,W].	interval 0 to W
mm	millimetres
C, C_1, C_2, C_3	constants
n, x, y	variables
i, j, k, l	subscripts
P, Q	digital image matrix of elements $p_{i,j}$, $q_{i,j}$
$P_1 \ldots P_n$	series of n image arrays
$\mathbf{F}$	array of functions $f_{i,j}$ such that $q_{i,j} = f_{i,j}(p_{i,j})$
$\left\{ \right.$	brackets
$p_{i,j} = p_{i,j,7} \times 2^7 + p_{i,j,6} \times 2^6 + \cdots + p_{i,j,0} \times 2^0$	decomposition of pixel into 8-bit binary word
$C = C_7 \times 2^7 + C_6 \times 2^6 + \ldots + C_0 \times 2^0$	decomposition of constant into 8-bit binary word
$q_{i,j} = f(p_{1,i,j}, p_{2,i,j})$	function of pixel at location i, j in images 1 and 2
$\boxed{\begin{matrix} n_{-1,-1} & n_{-1,0} & n_{-1,1} \\ n_{0,-1} & n_{0,0} & n_{0,1} \\ n_{1,-1} & n_{1,0} & n_{1,1} \end{matrix}}$	3 × 3 convolution kernel
$n_{0,0}$ etc.	kernel coefficients
$\sum_{k=-1}^{1} \sum_{l=-1}^{1} (\ )$	two-dimensional summation
$\lvert n_{k,l} \rvert$	modulus of $\lvert n_{k,l} \rvert$
$\sum_{k=-1}^{1} \sum_{l=-1}^{1} \lvert n_{k,l} \rvert$	summation of modulus
$\frac{\partial p}{\partial x}, \frac{\partial p}{\partial y}$	partial derivatives of p in x and y
$\theta(x, y)$	angular function
$G[p(x, y)]$	gradient function
$\sqrt{\left(\partial p / \partial x \right)^2 + \left(\partial p / \partial y \right)^2}$	root mean square
$\mathbf{N}$	pixel neighbourhood
$(n_1, n_2, n_3, \ldots n_N)$	series of pixel neighbourhood elements
R_k	k^{th} rank of series
$\boxed{\begin{matrix} m1 & m2 & m3 \\ m4 & m5 & m6 \\ m7 & m8 & m9 \end{matrix}}$	neighbourhood of binary image
{x1, x2, x3, ... xd}	feature vector of scalar properties

$$M_{xy} = \sum_i \sum_j i^x j^y p(i,j)$$ the moment of order $(x + y)$

$i' = M_{10}/M_{00}$
$j' = M_{01}/M_{00}$ centroid
μ_{xy} central moment

$$\mu_{xy} = \sum_i \sum_j (i - i')^x (j - j')^y p(i,j)$$

definition of central moment

n1	n2	n3
n4	n5	n6
n7	n8	n9

convolution kernel

$LTR(p_{i,j})$ Local Threshold Difference (LTD) operator
μm microns
X_1, X_2, X_3
Y_1, Y_2, Y_3
x_1, x_2, x_3
y_1, y_2, y_3
camera calibration points (micrometres)

$\begin{bmatrix} x_{robot} \\ y_{robot} \end{bmatrix}$ robot co-ordinates (micrometres)

$[x, y]$ local co-ordinates (micrometres)

$\begin{bmatrix} S_{0,0} & S_{0.1} \\ S_{1,0} & S_{1,1} \end{bmatrix}$ rotation and scale conversion matrix

where

$$S_{0,0} = \frac{X_2/y'_2 - X_3/y'_3}{x'_2/y'_2 - x'_3/y'_3}$$
$$S_{0,1} = \frac{Y_2/y'_2 - Y_3/y'_3}{x'_2/y'_2 - x'_3/y'_3}$$
$$S_{1,0} = \frac{X_2/x'_2 - X_3/x'_3}{y'_2/x'_2 - y'_3/x'_3}$$
$$S_{1,1} = \frac{Y_2/x'_2 - Y_3/x'_3}{y'_2/x'_2 - y'_3/x'_2}$$

and where

$x'_2 = x_2 - x_1$
$y'_2 = y_2 - y_1$
$x'_3 = x_3 - x_1$
$y'_3 = y_3 - y_1$

◆◆◆◆◆◆ I. INTRODUCTION

In this chapter we consider digital image processing techniques for colony and plaque counting, and automated picking. We also describe the practical implementation of the Flexys™ series of commercial products by Genomic Solutions (GSL) Ltd, Forge Close, St Neots, Cambs, PE19 3TP, U.K. (formerly PBA Technology Ltd).

A. Colony Picking

Every molecular biologist has picked colonies. Sterile toothpicks are often used to transfer relatively small numbers of clones, perhaps 12–24 for plasmid preparations. However, to prepare a genomic library, between 20 000 and 1 000 000 clones must be assembled. Without automation, this task would take months to complete. The Flexys™ colony and plaque picker will produce such a library, stored in an array of 96-well or 384-well microtiter plates, in a matter of hours.

B. Libraries

A library is usually prepared as the first stage of investigation into the genome of an organism. Extracted DNA is randomly fragmented and cloned into thousands of individual vectors, possibly plasmids, cosmids or phages. A host strain is then used to hold each of the recombinant clones. As the host replicates, the inserted DNA fragment is faithfully reproduced. *Escherichia coli* is commonly used to maintain plasmids and cosmids. Larger inserts of DNA are held by yeast strains. Sometimes a DNA fragment is inserted into a plaque such as the M13 virus, which is then permitted to infect a bacterial host. Whatever the chosen mechanism, local regions of identical colonies or plaques can be grown on an agar substrate, permitting the cloned DNA to be easily stored and retrieved. A sample of each colony or plaque is then carefully picked into a separate well of a microtitre plate to create a library of clones.

C. Colony and Plaque Picking

A colony or plaque region is considered suitable for picking on the basis of size, shape and colour. Colonies larger than a certain diameter may be rejected on the basis of age. Irregularly shaped regions are likely to be an amalgam of several neighbouring colonies that have merged as they have grown. Circular regions will probably have grown from a single host vector and contain copies of a single clone. It is possible to contrive that certain colonies or plaques are coloured red, brown, blue or are transparent to light. Blue growths indicate that no DNA has been inserted and regions of this colour will therefore be rejected for picking.

D. The Flexys™ Colony and Plaque Picker

The GSL Flexys™ colony and plaque picker is a sophisticated automated instrument designed to identify and pick colonies and plaques of varying morphology and colour (Figure 1). The Flexys™ uses a vision system to look for colonies on the surface of the input plates. A picking tool consisting of six (or 24) solenoid operated needles is used to sample the material and transfer it to a microtitre plate, sterilising between each cycle (Figure 2). Flexys™ is also able to generate gridded arrays at a range of densities on to nylon filters (Stewart *et al.*, 1995). This allows entire genomic libraries to be picked, gridded and replicated for large-scale screening purposes.

The basic layout of the Flexys™ carries up to eight GSL single-well rectangular agar plates but it can also be reconfigured to accept large 22 cm × 22 cm bioassay plates and 100 mm or 150 mm round Petri dishes. Up to eight output plates can be loaded at one time: either 96- or 384-well microtitre plates, or 96 × 2 ml deep-well boxes. A sterilisation fixture can be configured with up to three baths, including an ultrasonic bath, and a heater drying position. Genomic Solutions has recently announced an autoloader attachment that can stack and deliver up to 120 plates in a single unattended run.

Agar plates containing the colonies or plaques are loaded into a tray above a light box or transilluminator. The colonies or plaques to be picked are selected by an automated vision system consisting of a CCD camera connected to a frame grabber card in the controlling PC. The camera

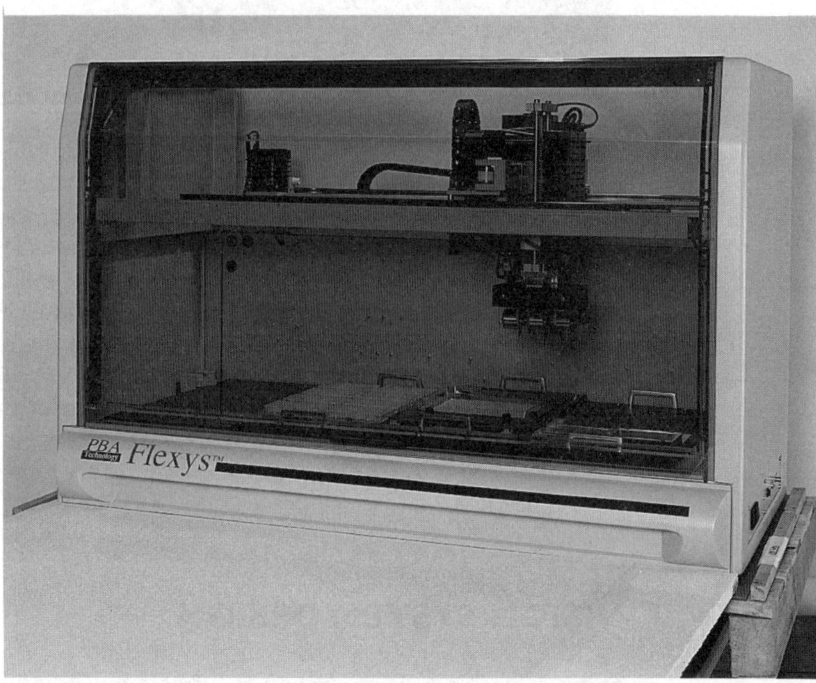

Figure 1. Flexys™ colony and plaque picker.

Figure 2. Six needle picking tool.

moves only in the x and y axes of the machine and has a resolution of 752 × 582 pixels with a field of view of 87 mm × 65 mm. The footprint at the agar surface is approximately 10 pixels per mm.

Before the Flexys™ begins the picking process it must first scan each plate containing colonies. The picture acquired from the camera is then analysed by a digital image processing algorithm. Not all colonies are suitable to be picked. Parameters can be set by the user that enable the machine to select colonies or plaques based on size, shape and colour or density. The x, y co-ordinates of each colony centre are then used to target the picking needles.

◆◆◆◆◆◆ II. VISION SYSTEM DESIGN

Designing a machine vision system requires a multidisciplinary mix of skills and technologies including mechanical engineering, illumination and optics, image formation and sensing, analog and digital electronics, computer science and a pot-pourri of image processing algorithms. Some of the considerations in the design process leading to each element of the vision system are described below, beginning with illumination and optics.

A. Illumination Techniques

It is worth while spending time and effort investigating lighting and viewing configurations. The physical illumination of an object under inspection is in effect the first signal conditioning stage of an imaging system. Appropriate illumination can enormously ease the subsequent image processing workload, increasing the reliability of the system and reducing its cost. For an extensive treatment, Batchelor *et al.* (1985) illustrate over sixty useful illumination and viewing techniques.

The Flexys™ Picker is required to inspect colony or plaque growth on a translucent agar substrate contained in a transparent Petri dish. Immediately this suggests back lighting (or transillumination) of some sort. A standard light box, of the type used for viewing X-radiographs, can be employed. Such a light box consists of a light source (often a pair of fluorescent tubes), a back reflector and a scattering screen to give an even surface of diffuse light.

When a Petri dish is placed over the light box, colonies show up as dark circles because they absorb transmitted light. Under the same conditions, however, plaques are difficult to see at all. Colonies consist of protruding globules that diffuse and scatter incident light, and grow on a clear substrate. For plaques, the situation is inverted: plaques consist of clear depressions that erode into a diffusing background lawn of bacteria. We have found that the illumination conditions required to acquire high contrast images of plaques are entirely different to the set-up needed for colonies.

1. Colony illumination

Back lighting alone is not helpful if colour or density discrimination is required: colonies simply show up as dark circular masses. An immediate improvement can be made with dark field illumination. The simplest way to achieve dark field illumination is to put a black mat between the light box and the Petri dish, such that the lighting comes from the sides (Figure 3). Looking from above, there is no direct light in the field of view, giving a black background. When the indirect side light encounters imperfections or diffusing areas, sufficient energy is scattered towards the viewing direction to give a high contrast image. Dark field illumination is especially useful for looking at clear objects such as glassware.

With dark field illumination, the contrast between blue and white colonies is easily distinguishable. Diffusion of light by the agar gel gives a background illumination that varies slowly across the field of view, with occasional dust blemishes. These effects do not impact significantly on the image processing performance.

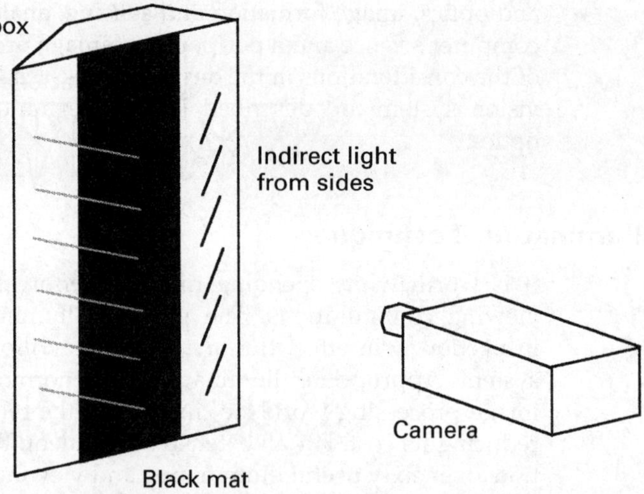

Figure 3. Dark field illumination.

2. Plaque illumination

Clear plaques can hardly be seen using back lighting, although blue or brown plaques show up as dark zones in the final image. If dark field illumination is used, all plaque areas become dim and it is difficult to distinguish between clear and blue regions. To understand this, consider the situation with diffuse back lighting.

(a) Diffuse back lighting

Scattered light rays from the transilluminator strike the underside of the Petri dish at a range of angles, and pass almost unimpeded through clear plaques. Looking from above a clear region, only a fraction of the light energy is directed into the field of view, while the rest escapes to the sides (Figure 4a).

The background lawn of bacteria acts as a second diffusing screen. A light ray entering this diffusing area is attenuated slightly and scattered. Looking at a small point above the lawn, the energy is an average of the light at that point, plus that of its neighbourhood. Light rays leaving the diffusing lawn are scattered at various angles, again leaving only a few in the viewing direction (Figure 4b). Therefore the proportion of light energy received in the direction of view from the clear region is similar to the amount from the diffusing area. What small contrast exists is due to the attenuation of light through the diffusing lawn.

(b) Parallel back lighting

An enormous improvement can be achieved if parallel back lighting is used, and observed from directly above. Here, all the incident light is in

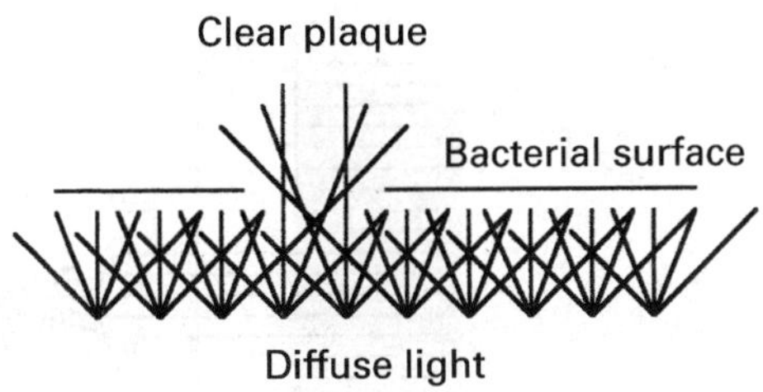

(a)

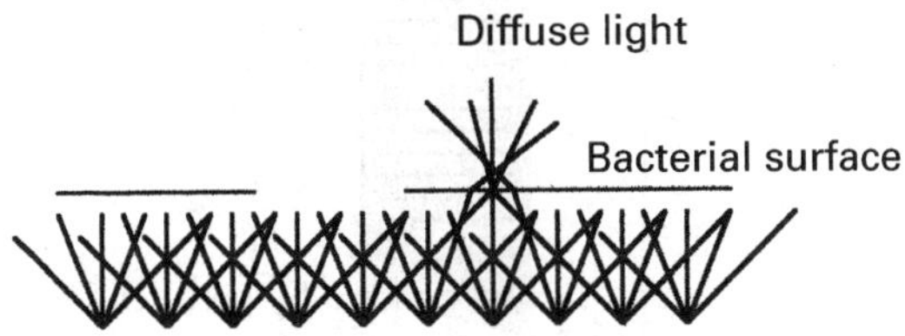

(b)

Figure 4. (a) Illuminating plaques with diffuse back lighting; (b) illuminating bacterial surface with diffuse back lighting.

the viewing direction. As the light passes through a plaque region, the energy is concentrated in the field of view (Figure 5a).

However, when the parallel light strikes the diffusing lawn, the rays are scattered in all directions. The amount of energy directed towards the camera is therefore reduced (Figure 5b). Parallel back lighting therefore gives high contrast. Clear plaques appear as bright regions. Any coloration or staining can readily be seen.

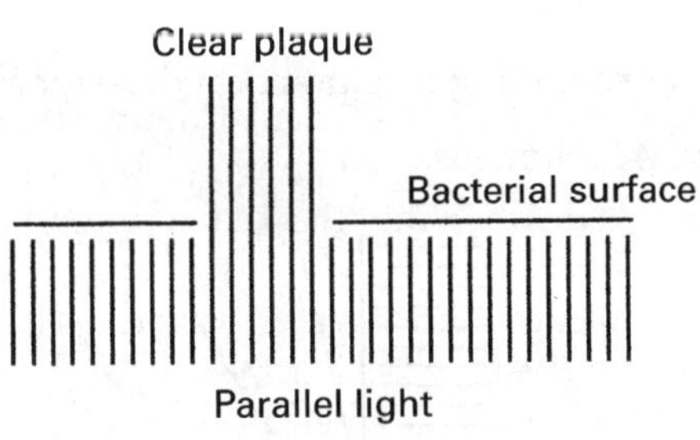

(a)

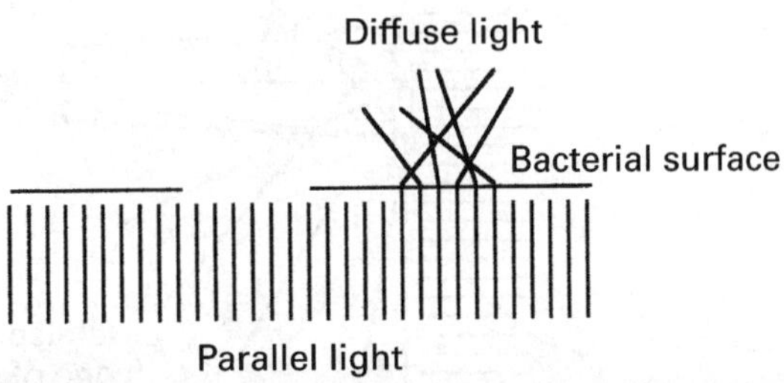

(b)

Figure 5. (a) Illuminating plaques with parallel back lighting; (b) illuminating bacterial surface with parallel back lighting.

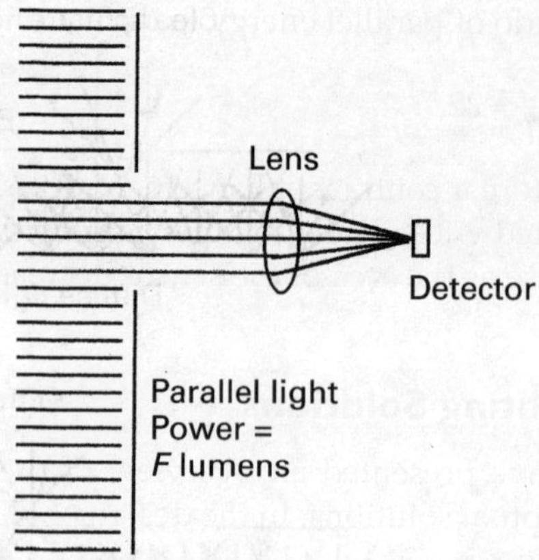

Figure 6. Luminous flux emitted by parallel light.

An indication of the magnitude of the contrast difference can be seen if a small region of the parallel light illuminator is considered (Figure 6). With parallel light, the luminous flux (visible power, *F*) emitted from this region is the same as that received at the detector.

If an ideal diffusing screen is placed in the path of the parallel beam, the energy is uniformly directed into a hemispherical volume (Figure 7). At a distance, *r*, from the source, the luminous flux, *F*, is distributed over the area of a hemispherical cap of radius *r*. The illumination, *E* is then:

$$E = \frac{F}{2\pi r^2} \text{ lux}$$

For the Flexys™ , where $r = 340$ mm:

$E = 1.38F$ lux

The power falling on a practical detector area of 32 mm square is:

$E = 0.044\ F$ lumens.

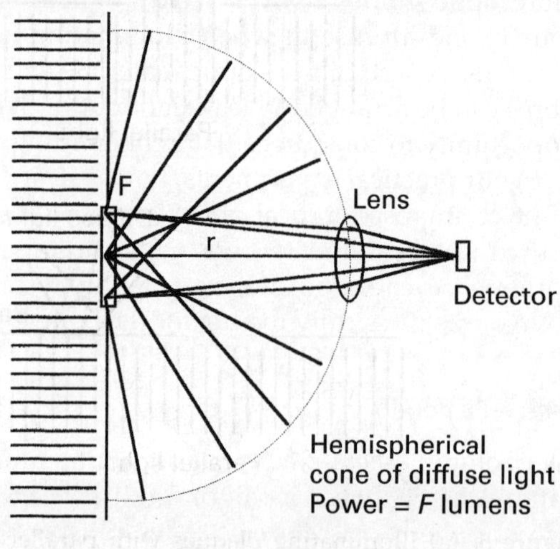

Figure 7. Luminous flux distributed by diffuse light.

The ratio of parallel energy to diffuse energy is:

$$\frac{F}{0.044F} = 22.7$$

Therefore a contrast enhancement of over an order of magnitude can be expected with ideally parallel light, with no other losses. This figure will increase as the area of the sensor decreases.

B. Practical Lighting Solutions

We have presented an overview of the problems of lighting, and their conceptual solutions. In the design of Flexys™ , practical solutions had to be devised.

1. Transilluminator

A slimline light box was designed, housing a pair of fluorescent tubes, a reflector and a diffusing perspex screen. A low voltage electronic ballast is used to drive the fluorescent tubes at 100 kHz to avoid stroboscopic effects at the camera.

2. Light emitting surfaces

Fluorescent tubes and their associated drive circuitry take up space. In the case of Flexys™ there is sufficient drop below the bed of the robot to house the light box enclosure. In future systems we might not have this luxury. We investigated two alternatives: fibre optic mats and electroluminescent panels.

(a) Fibre optic mats

Fibre optic guides deliver "cold" light from an incandescent or other source, and are useful when illuminating heat sensitive or inaccessible locations. A web of fibre optic material can be netted into a flat mat. The fibres can be arranged to leak light out of the planar surface, offering the opportunity to construct a flat light emitting surface.

In our practical experiments, however, we found that the coarseness of the web imposed a local grid pattern on the image. Also, a 150 W light source was required to give a useful emission intensity from the mat. Although it could be sited remotely, the standard housing for the light source was physically too big for the boundaries of the robot's casing.

(b) Electroluminescent panels

Electroluminescent panels consist of thin films of light emitting phosphor sandwiched between a pair of conductive electrodes. An alternating voltage is applied across the electrodes, causing light to be emitted during each half-cycle. The luminance of the generated light increases with the applied voltage and frequency which varies between about 40 and 220 volts (AC) and 50 Hz and 5 kHz. Electroluminescent panels have a working lifetime of over 10 000 hours during which the output luminance gradually diminishes. Higher values of applied voltage reduce the working life.

The advantages of electroluminescent panels is that they have low weight, conserve space, can be produced in a wide variety of planar and curved forms, degrade gracefully rather than fail catastrophically, and consume low power.

Thin electroluminescent panels are made to specification by the Quantex Corporation, 1 Research Court, Rockville, Maryland 20850-3221. The weight of a typical Quantex electroluminescent panel is 0.1 g cm^{-2} (0.001 lb in^{-2}) with a thickness of 0.5 mm (0.020 in). Current consumption is approximately 0.14 mA cm^{-2} (0.9 mA cm^{-2}) at a voltage of 115 V, 400 Hz.

(c) Light emitting polymers

Recent developments in molecular electronics have produced polymer semiconductors that can emit light at colours ranging from deep blue to the near infrared. In addition to the advantages of electroluminescent phosphors, light emitting polymers operate at low DC voltages (3 V). This technology is being developed by Cambridge Display Technology, 181a Huntingdon Road, Cambridge, CB3 0DJ, UK.

3. Dark field illumination

As mentioned earlier, the simplest way of producing dark field illumination is to put a narrow black mat on the surface of the light box. Light illuminates the object from the sides; there is no direct path to the camera lens. With this approach, the major problem is that the illumination intensity varies considerably across the field of view. Less light is scattered at the centre of the mat than at its sides. With Petri and bioassay plates there are problems of glare and glinting at the edges of the light box. An alternative is to use a louvred grille to distribute the light more evenly as shown in Figure 8.

4. Parallel back light

To illuminate plaque regions, a uniform source of parallel back lighting is required. Two alternatives were examined: parabolic reflectors and the use of a special material known as *brightness enhancement film*.

(a) Parabolic reflectors

Parallel light can be produced with a parabolic reflector with a light source placed at its focus. We experimented with a double reflector housing a pair of fluorescent tubes.

In practice we found that uniform illumination was difficult to obtain, partly because of irregularities in the reflective surface, and partly

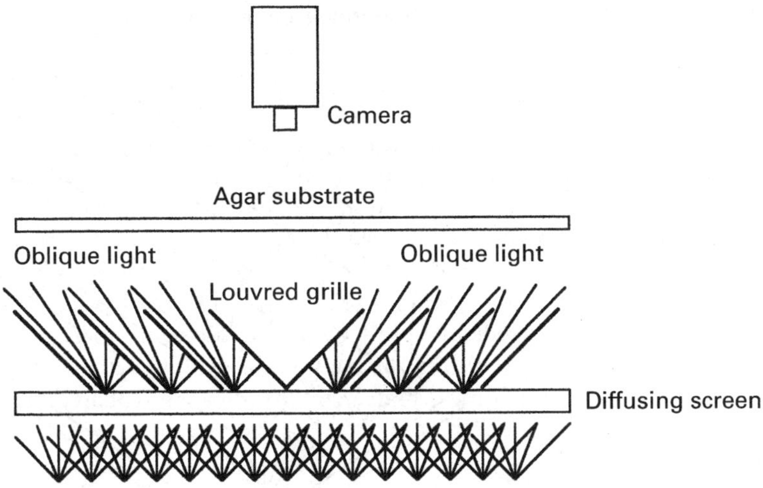

Figure 8. Dark field illumination using louvred grille.

because the intensity of the direct light from the tubes was significantly brighter than the reflected background, giving rise to two bright strips along the length of the reflector.

(b) Brightness enhancement film

An alternative to parabolic reflectors for parallel light generation is a material known as brightness enhancement film, produced by 3M Electronic Display Lighting (3M Center, Building 225-4N-14, St Paul, MN 55144-1000, USA). Designed to improve backlight efficiency in laptop computers, instrumentation and other displays, the film uses a structure of microprisms to enhance light intensity. Placed over a diffusing surface, the film employs a mixture of refraction and internal reflection to funnel the diffuse light into a fan of up to 70° (Figure 9).

Two films placed at orthogonal angles produce an approximate cone of light. The illumination is sufficiently close to parallel to give a significant contrast improvement over a diffuse transilluminator when imaging plaques. The lighting generated is completely uniform.

When viewed normal to the direction of illumination, the maximum contrast improvement over a uniformly diffusing surface is given by the ratio of the illumination of the spherical cap of the diffusion cone, to that of the entire hemisphere (Figure 10).

For a hemisphere:

$$E_1 = \frac{F}{2\pi r^2} \text{ lux}$$

For the spherical cap:

$$E_2 = \frac{F}{A} \text{ lux}$$

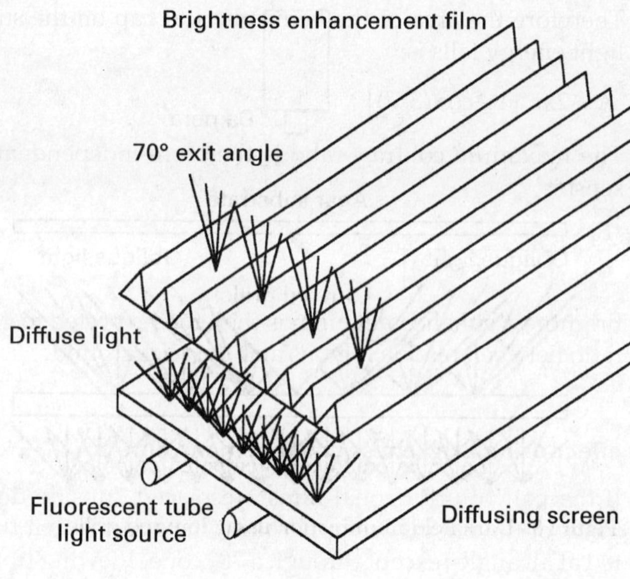

Figure 9. Brightness enhancement film.

where A is the area of a spherical cap, given by:

$$A = 2\pi \int_a^r r \, dx$$

or

$$A = 2\pi r [x]_a^r$$

A diffusion cone of 70° is produced by the brightness enhancement film, therefore the lower limit of the integral is:

$a = r \cos(70°/2) = r \cos(35°)$

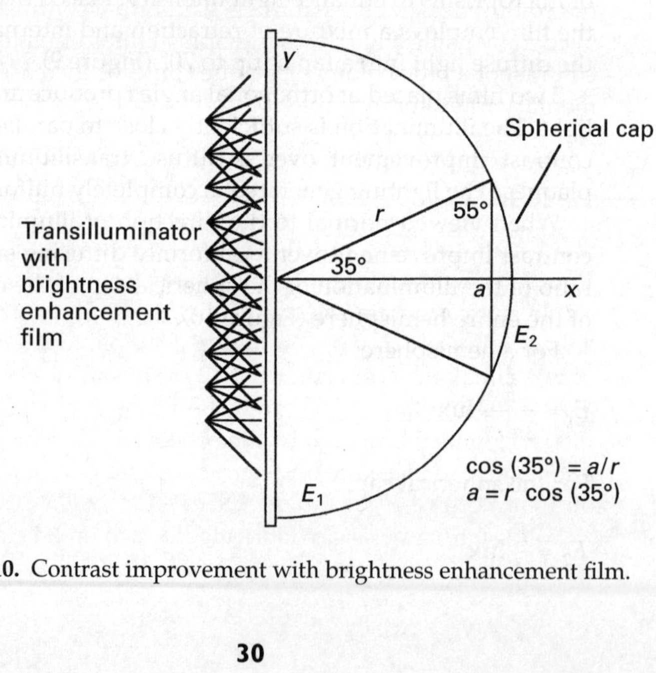

Figure 10. Contrast improvement with brightness enhancement film.

Therefore the area of the hemispherical cap on the surface of which the light energy falls is:

$$A = 2\pi r^2\left[1 - \cos(35°)\right]$$

The maximum contrast ratio is therefore independent of the area of the sensor:

$$\frac{E_1}{E_2} = \frac{1}{\left[1 - \cos(35°)\right]} = 5.5$$

Brightness enhancement film is therefore expected to give a 550% contrast ratio between plaque regions and the background.

5. Dark field effect using brightness enhancement film

If the pair of orthogonal films are placed "upside down" such that the bases of the prisms are uppermost, the effect is that the light is transmitted at all angles except through a 70° cone. Viewing from above, therefore, no direct light is observed. This is an almost ideal dark field configuration. Because the surface of the film behaves as a reflector, the contrast of the dark field image is reduced in comparison to that obtained with a matt black surface. The illumination is highly uniform over the field of view, however.

In the final product brightness enhancement film was used to give both dark field illumination and parallel lighting, simply by reversing the film.

C. Camera

Although discrimination between blue/clear or red/clear regions is required, a monochrome camera, rather than colour, can be used because the coloured regions are darker. Additional contrast could be obtained if needed by using a narrow band optical filter fitted to the camera lens. The contrast of a blue colony or plaque could be increased with a red complementary filter, for example.

1. Image sensor technology

Solid state cameras are the natural choice for a robot vision system. Compared with vacuum-tube cameras, semiconductor devices are more stable, more accurate and more reliable. Semiconductor cameras are available as a linear array of commonly 2048 photodetector sites, or as a two-dimensional array of typically 512 × 512 sensor elements. Resolutions of 1320 × 1035 and upwards can be found in commercial devices. High geometric stability is ensured by the process of semiconductor fabrication; the photosites are placed to an accuracy of one part in ten thousand, typically on 10 μm centres.

Solid state cameras are fabricated in two broad classes: as charge transfer devices (CTDs) or as a linear photodiode array (LPA).

(a) Charge transfer devices

Sensors based on charge transfer devices use incident light to generate charge carriers. The charge depends on the illumination intensity and duration (known as the integration time). These devices are highly linear in terms of their electrical response to light intensity. High photosite densities can be fabricated per unit area of semiconductor wafer.

However, the active area of each site is produced by local electrostatic field effects that are not constant across the array. These irregularities cause a fixed background *pattern noise*. In addition to this, thermally generated charge carriers cause a time-varying background signal (or dark current) even in the absence of light. This offset is a linear function of integration time, but it is highly sensitive to temperature, doubling for every 6–8°C increase. Often, the on-chip readout circuitry is the dominant source of thermally generated noise.

(i) CCDs and CIDs

There are two important categories of CTD device: charge coupled devices (CCDs) and charge injection devices (CIDs). CCDs and CIDs differ in the way the charge is read from the array of sensors. CCD arrays actually shift the packets of accumulated charge to a single sensing electrode that converts it to a voltage. CIDs use a separate sensing electrode for every photosite, and every element of the array can be addressed individually. CCDs are prone to an effect known as *blooming* where charge from one photosite overspills into its neighbours. They have greater readout noise than CIDs. CIDs can be radiation hardened making them suitable for UV or X-ray imaging or operation in hostile environments. Because CIDs leave the charge intact, the image can be monitored in real-time as it builds up. CIDs, though, are less sensitive to light than CCD cameras. Most commercially available cameras are based on CCDs.

(b) Linear photodiode arrays

Linear photodiode arrays consist of individual diffused PN junctions, together with associated readout circuitry. The PN junctions are reversed-biased allowing charge to accumulate in the depletion region. Carriers are generated under illumination, permitting the charge to be conducted out. In this case the remaining charge depends on the illumination intensity and the discharge time. When the diode is recharged, there is a spike of current that is related to the light intensity. Photodiode arrays are not susceptible to blooming, and are far more uniform that CTDs, because the photosites are not created by field effects. A large quantity of readout circuitry is required that limits commercial photodiode arrays to the one-dimensional linescan format.

(c) CCD temperature problems

We have found that, in practice, thermal noise can cause problems. During our initial experiments with plaque picking where we had a low contrast image, we found that the plaque detection count deteriorated with time, to the extent that after about an hour we were not identifying any plaques at all. It turned out that the interior of the robot's enclosure was getting warm because part of the sterilisation cycle uses an electrical heater. The increase in temperature caused enough thermal carriers to disrupt the image by imposing a small amount of granular "salt-and-pepper" noise.

If the original image had had higher optical contrast, the noise component would not have been significant. However, in this case the additive noise was enough to cause the detected edges of the plaque regions to take on a ragged appearance, ruining their circularity and causing the image processing algorithm to reject them. We approached the problem by changing the lighting conditions as described earlier. We also looked at several different cameras, and found that the TM6-CN CCD camera by Pulnix Europe Ltd (Pulnix House, Aviary Court, Wade Road, Basingstoke, Hampshire RG24 8PE, UK) gave a reasonable immunity to thermal noise, with a specified operating temperature range of –10°C to +50°C.

The TM-6 CCD camera produces a CCIR format video signal (625 lines at 50 Hz) at 1 Vp-p. The spatial resolution is 752 pixels horizontally and 582 vertically, imaged on to a CCD chip of sides 6.4 mm × 4.8 mm. At the chip surface the photosite dimensions are 8.6 μm × 8.3 μm. The physical dimensions of the camera body are 45 mm(w) × 39 mm(h) × 92 mm(d).

D. Lens Selection

Standard PBA Technology rectangular Petri plates are imaged in three overlapping sections. The horizontal axis of the camera is aligned with the width of the Petri plate, which measures 83 mm. To select the lens to be used, the relationship between its focal length and its position from the object plane must be considered. This relationship is:

$$f = 6.4\frac{L}{H}$$

and

$$f = 4.8\frac{L}{V}$$

where f is the focal length of the lens, L is its position above the object plane, H is its footprint projected horizontally (with respect to the camera CCD) on to the object plane, and V is its similar footprint projected in the CCD vertical direction.

In our case we required a horizontal footprint of $H = 87$ mm, giving a margin of 2 mm on either side of the Petri plate. Selecting a fixed 25 mm focal length lens by Cosmicar-Pentax (Pentax Corporation, 35 Inverness Drive East, PO Box 6509, Englewood, Colorado 80155-6509, USA) the above relationships produce:

$$L = f\frac{H}{6.4} = \frac{25\,\text{mm} \times 87\,\text{mm}}{6.4} = 340\,\text{mm}$$

$$V = 4.8\frac{L}{f} = 4.8\frac{340\,\text{mm}}{25\,\text{mm}} = 65\,\text{mm}$$

The camera was therefore mounted such that its lens is positioned at a distance of 340 mm above the object surface.

Since the TM-6CN has 752 photosites horizontally, the footprint projected by each pixel on to the object surface in that direction is:

$$l = \frac{87\,\text{mm}}{752} = 0.116\,\text{mm}$$

Similarly the vertical footprint per pixel is:

$$v = \frac{65\,\text{mm}}{582} = 0.112\,\text{mm}$$

◆◆◆◆◆◆ III. DIGITAL IMAGES

The image detected by the monochrome camera can be represented as a two-dimensional continuous function $p(x, y)$ denoting the intensity p at any point (x, y). The brightness range that can be handled by the sensor is called the grey scale, in which p lies between 0 and W:

$$0 \le p \le W$$

where $p = 0$ is defined as black, and $p = W$ is full white. Between these two extreme values, p is a continuous variable representing a darker or lighter shade of grey.

A. Image Sampling

Digital computers deal with discrete digital quantities rather than continuous functions. To convert the function $p(x, y)$ into a form suitable for digital processing it can be sampled as a two-dimensional array of discrete integers. Each element of this digital image is known as a pixel. The dimension $I \times J$ of the array is the spatial resolution of the image. Each pixel $p_{i,j}$ lies in the grey scale range $[0, W]$.

In computing it is often convenient to represent the range of values taken by I, J and W by integer powers of two. Typical image resolutions are 512 × 512 or 1024 × 1024 with an 8- or 16-bit grey scale.

Binary images are a useful case where $W = 1$. Each pixel therefore only has two possible values usually representing an object and its background. Digital images can also represent colour and multispectral components, three-dimensional depth, and motion.

B. Analog-to-Digital Conversion

Electronic hardware is required to convert the standard CCIR signal, supplied by the camera, into its digital representation. There are many commercial plug-in cards known as frame grabbers, available for image digitisation. Commercial frame grabbers have various combinations of on-board memory, processing capability and real-time display using a dedicated external monitor.

For the Flexys™ series, the Data Translation DT3155 frame grabber was selected. The DT3155 has a spatial resolution of 768 × 576 and digitises to 8 bits giving a 0–255 grey level interval. High speed analog circuitry is used to prevent loss of image sharpness at extreme intensity transitions. Sampling jitter is specified at no more than ±5 ns. There is some real-time processing capability on board, consisting of analog and digital contrast adjustment, spatial scaling and clipping.

Real-time video can be displayed on the host PC display, useful for adjusting the lens focus and f-number. No external video monitor is required because the PCI bus is used to transfer the image into PC memory at 45 Mb s^{-1} or higher. For the same reason, the DT3155 does not require on-board memory.

At the framestore, the projected footprint at the object plane is:

$$l = \frac{87\,\text{mm}}{768} = 0.113\,\text{mm}$$

Similarly the vertical footprint is:

$$v = \frac{65\,\text{mm}}{576} = 0.113\,\text{mm}$$

The horizontal resolution is determined by the 752 pixels per line available at the camera.

◆◆◆◆◆◆ IV. DIGITAL IMAGE PROCESSING

An enormous body of image processing techniques has been developed since the 1960s. An extensive coverage is given by Castleman (1996), Gonzalez and Wintz (1987), Pratt (1991) and Rosenfeld and Kak (1982). Although there is no overall unifying theory, in general an image processing application can be subdivided into low-level operations, intermediate-level processing and high-level processing.

Low-level operations enhance or isolate specific features of the original image, for example edges, surfaces, regions, complete objects or groups. Operations at this stage are performed at the pixel level, the output consisting of a set of *iconic* images that preserve the spatial relationships of the original features. At the subsequent stage of intermediate-level processing, the iconic images are integrated into a descriptive or symbolic form. The important characteristic of intermediate-level processing is that the pictorial information is reduced to a set of descriptors describing the essential image features. The quantity of data is enormously reduced. High-level processing interprets the symbolic descriptions, and appropriate action is initiated. In advanced systems, high-level tasks may involve artificial intelligence techniques such as predicate logic, planning and model matching.

A. Low-level Operations

Operating at the pixel level, these processes are arithmetic (analytic or non-linear) or logical functions defined over the entire space of the image array. Detailed descriptions of many such operations are found in Batchelor and Whelan (1997), and Bassmann and Besslich (1995).

Low-level operations can be subclassified into the categories of pre-processing, segmentation, post-processing and labelling. Pre-processing operations standardise the original image prior to further work. Poor signal to noise ratio, and low contrast are typically compensated for at this stage if required. Segmentation operations isolate specified components in the image. Distinct objects and regions and their properties are segmented typically on the basis of intensity, edges, texture and colour. Post-processing consolidates the segmentation process by integrating anomalies such as incomplete edges, regions or isolated points. Labelling associates the individual pixels in the post-processed image with a particular region.

A low-level operation accepts one or more input images $P_1 \ldots P_n$ as arguments, outputting a single image, *Q*. The function, *f*, to be applied to the image may be isotropic or anisotropic.

1. Isotropic functions

An isotropic function is spatially invariant. In effect, the same function is applied independently to each pixel in the image for all *i*, *j* in the domain of *f*:

$$Q = f(P)$$

where *f* is applied with respect to the $p_{i,j}$ of the input image, and yields a point, $q_{i,j}$.

2. Anisotropic functions

Although the majority of low-level functions are isotropic, there is also a class of anisotropic functions where the effective operation depends on the spatial location of the image. In general therefore:

$$Q = \mathbf{F}(P)$$

where **F** is an array of functions $f_{i,j}$ such that:

$$q_{i,j} = f_{i,j}\left(p_{i,j}\right)$$

3. Point transformations

Point transformations are one-to-one pixel mappings where each $q_{i,j}$ is derived from the corresponding $p_{i,j}$ in each of one or more input images. There is a direct relationship between the value of the input and that of the output pixel.

An improvement in signal-to-noise ratio can be made, for example, by frame averaging: summing the corresponding pixels from n images, and dividing the sum by the number of images:

$$q_{i,j} = \frac{\left(p_{1,i,j} + p_{2,i,j} + \cdots + p_{n,i,j}\right)}{n}$$

Despite the above example, most low-level operations are either monadic (a single input image) or dyadic (a pair of input images).

(a) Monadic point transformations

Monadic point operators process each point $p_{i,j}$ in the original image using a transformation f, producing a point q at the same location i, j in the output image:

$$q_{i,j} = f\left(p_{i,j}\right)$$

In practice, monadic point operations are readily performed at video rates with look-up tables (LUTs). Virtually all commercial frame grabbers have on-board RAMs that are addressed directly by the binary digits representing the grey level pixel values. The content of the RAM is loaded with the transformation. As the value of each pixel is presented to the RAM address lines, the output of the RAM will give the transformed values. Any arbitrary mapping, linear or non-linear, can be loaded into the LUT. Examples of monadic operators include contrast manipulation, image negation and binary thresholding.

(i) Add Constant

A constant C is added to the value of each pixel:

$$q_{i,j} = p_{i,j} + C$$

In practice it is important to avoid problems of numerical overflow caused if the value generated is outside the permitted grey level range $0 \le q_{i,j} \le W$. If the result is less than zero (if C is negative and its modulus is greater than $p_{i,j}$) we therefore clamp the output at zero. Similarly the output is clamped at maximum white, W, if the result is greater than white:

$$q_{i,j} = \begin{cases} 0, & \text{if } \left(p_{i,j} + C\right) < 0 \\ p_{i,j} + C, & \text{if } 0 \le \left(p_{i,j} + C\right) \le W \\ W, & \text{if } \left(p_{i,j} + C\right) > W \end{cases}$$

(ii) Multiply by Constant

An improvement in contrast can be obtained by multiplying each $p_{i,j}$ by a constant, C:

$$q_{i,j} = p_{i,j} \times C$$

Introducing clamping to avoid overflow:

$$q_{i,j} = \begin{cases} 0, & \text{if} \quad \left(p_{i,j} \times C\right) < 0 \\ p_{i,j} \times C, & \text{if} \quad 0 \le \left(p_{i,j} \times C\right) \le W \\ W, & \text{if} \quad \left(p_{i,j} \times C\right) > W \end{cases}$$

(iii) Divide by Constant

Each $p_{i,j}$ is divided by a constant, C:

$$q_{i,j} = \frac{p_{i,j}}{C}$$

With clamping:

$$q_{i,j} = \begin{cases} 0, & \text{if} \quad \frac{p_{i,j}}{C} < 0 \\ \frac{p_{i,j}}{C}, & \text{if} \quad 0 \le \frac{p_{i,j}}{C} \le W \\ W, & \text{if} \quad C = 0 \quad \text{or} \quad \frac{p_{i,j}}{C} > W \end{cases}$$

(iv) Negation

An effect similar to that of a photographic negative can be produced by subtracting the value of each pixel from maximum white:

$$q_{i,j} = W - p_{i,j}$$

No clamping is needed because the result is always contained in the range

$0 \le q_{i,j} \le W$.

(v) Intensity Squaring

An effective non-linear contrast enhancement function is obtained by squaring the pixel values. The output is normalised to the permitted grey level range by dividing by maximum white:

$$q_{i,j} = \frac{p_{i,j}^2}{W}$$

(vi) Gamma Correction

Generalising the intensity squaring operation, a gamma correction function can be produced:

$$q_{i,j} = \frac{p_{i,j}^n}{W^{(n-1)}}$$

(vii) Highlight Intensities

In some instances, features of interest occupy a distinct range of grey levels that can be highlighted or thresholded out. Highlighting creates regions of constant intensity, C, where the input pixels have grey level values between C_1 and C_2:

$$q_{i,j} = \begin{cases} C, & \text{if} \quad C_1 \le p_{i,j} \le C_2 \\ 0, & \text{if} \quad 0 \le p_{i,j} < C_1 \quad \text{or} \quad C_2 < p_{i,j} \le W \end{cases}$$

(viii) Intensity Threshold

This non-linear operation is important because the output takes on only two possible values: 0 and C_3, and is therefore known as a binary image:

$$q_{i,j} = \begin{cases} C_3, & \text{if} \quad C_1 \le p_{i,j} \le C_2 \\ 0, & \text{if} \quad 0 \le p_{i,j} < C_1 \quad \text{or} \quad C_2 < p_{i,j} \le W \end{cases}$$

C_1 and C_2 are two grey level values within which the intensity of the region of interest lies. C_3 is often set to full white, W, for purposes of display. Binary images are often generated as a result of the process of segmenting salient features. There is a large class of binary image processing operators used for shape identification and measurement that will be introduced later.

(ix) Bitwise Logical Operations

Representing the pixel values in binary notation, operations can be performed between the corresponding bits of $p_{i,j}$ and a constant, C. Typically, $p_{i,j}$ is an 8-bit binary word:

$$p_{i,j} = p_{i,j,7} \times 2^7 + p_{i,j,6} \times 2^6 + \cdots + p_{i,j,0} \times 2^0$$

and C is similarly:

$$C = C_7 \times 2^7 + C_6 \times 2^6 + \cdots + C_0 \times 2^0$$

Logical AND

$$q_{i,j} = \left(p_{i,j,7}\,\text{AND}\,C_7\right) \times 2^7 + \left(p_{i,j,6}\,\text{AND}\,C_6\right) \times 2^6 + \cdots + \left(p_{i,j,0}\,\text{AND}\,C_0\right) \times 2^0$$

Assuming 8-bit pixels, setting C to binary 1110000 will truncate the brightness to only eight possible grey-level values, introducing an artificial contouring of the image.

Logical OR

$$q_{i,j} = \left(p_{i,j,7}\,\text{OR}\,C_7\right) \times 2^7 + \left(p_{i,j,6}\,\text{OR}\,C_6\right) \times 2^6 + \cdots + \left(p_{i,j,0}\,\text{OR}\,C_0\right) \times 2^0$$

C is used as a mask to set selected bits of $p_{i,j}$.

Logical Exclusive OR

$$q_{i,j} = \left(p_{i,j,7}\,\text{EXOR}\,C_7\right) \times 2^7 + \left(p_{i,j,6}\,\text{EXOR}\,C_6\right) \times 2^6 + \cdots + \left(p_{i,j,0}\,\text{EXOR}\,C_0\right) \times 2^0$$

C is used to complement selected bits of $p_{i,j}$.

(b) Dyadic point transformations

Dyadic point transformations are point-by-point operations from two input images, P_1 and P_2, producing an output image Q:

$$q_{i,j} = f\left(p_{1,i,j}, p_{2,i,j}\right)$$

(i) Image Addition

Adding two images on a point-by-point basis gives the familiar effect of photographic double exposure:

$$q_{i,j} = \frac{(p_{1,i,j} + p_{2,i,j})}{2}$$

(ii) Image Subtraction

Spatial differences between images are obtained by subtraction. If the two images are of a moving object on an invariant background, the leading and trailing edges can be extracted:

$$q_{i,j} = \frac{(p_{1,i,j} - p_{2,i,j})}{2} + \frac{W}{2}$$

If there is no difference between the input images, the output will be half-white, $W/2$. Interestingly, if image P_1 is a blurred (or low-pass filtered) version of P_2, then subtraction gives regions where the grey level is changing rapidly across the image, often corresponding to edges in an object under inspection.

In some applications the background lighting may vary over the area, or there may be spatial non-uniformities in the imaging device that impose non-uniform variations across the field of view. Attempts to improve the contrast of such images may also enhance any features of the background. A straightforward way of eliminating such effects is to subtract an image of the background only, from an image of the object on the same background. It is important to keep the camera fixed with respect to the background so that the images are kept in registration.

(iii) Image Multiplication

Dyadic image multiplication is the product of corresponding pixel values, normalised by the scaling factor, W:

$$q_{i,j} = \frac{(p_{1,i,j} \times p_{2,i,j})}{W}$$

Often, P_2 is a spatial mask of known or specified characteristics.

(iv) Image Division

An alternative to background subtraction, image division is used to normalise the output when there are fixed non-uniformities, possibly caused by defects in the camera or other sensor:

$$q_{i,j} = \frac{p_{1,i,j}}{p_{2,i,j}}$$

Clamped to:

$$q_{i,j} = \begin{cases} 0, & \text{if } p_{1,i,j} = 0 \text{ and } p_{2,i,j} = 0 \\ \dfrac{p_{1,i,j}}{p_{2,i,j}}, & \text{if } 0 \le \dfrac{p_{1,i,j}}{p_{2,i,j}} \le W \\ W, & \text{if } p_{1,i,j} > 0 \text{ and } p_{2,i,j} = 0 \end{cases}$$

(v) Dyadic Maximum

A non-linear operation, this is the point-by-point maximum of two images:

$$q_{1,j} = \begin{cases} p_{1,i,j}, & \text{if} \quad p_{1,i,j} \geq p_{2,i,j} \\ p_{2,i,j}, & \text{if} \quad p_{2,i,j} > p_{1,i,j} \end{cases}$$

(vi) Dyadic Minimum

The point-by-point minimum of two images:

$$q_{i,j} = \begin{cases} p_{1,i,j}, & \text{if} \quad p_{1,i,j} \leq p_{2,i,j} \\ p_{2,i,j}, & \text{if} \quad p_{2,i,j} < p_{1,i,j} \end{cases}$$

(vii) Dyadic Bitwise Logical Operations

Representing the pixel values in binary notation, operations can be performed between the corresponding bits of $p_{1,i,j}$ and $p_{2,i,j}$.

Logical AND

$$q_{i,j} = \left(p_{1,i,j,7} \text{ AND } p_{2,i,j,7}\right) \times 2^7 + \left(p_{1,i,j,6} \text{ AND } p_{2,i,j,6}\right) \times 2^6 + \cdots + \left(p_{1,i,j,0} \text{ AND } p_{2,i,j,0}\right) \times 2^0$$

When performed on binary images, ANDing gives the intersection of regions, and is equivalent to the dyadic minimum operation.

Logical OR

$$q_{i,j} = \left(p_{1,i,j,7} \text{ OR } p_{2,i,j,7}\right) \times 2^7 + \left(p_{1,i,j,6} \text{ OR } p_{2,i,j,6}\right) \times 2^6 + \cdots + \left(p_{1,i,j,0} \text{ OR } p_{2,i,j,0}\right) \times 2^0$$

When performed on binary images, ORing gives the union of regions, and is equivalent to the dyadic maximum.

Logical Exclusive OR

$$q_{i,j} = \left(p_{1,i,j,7} \text{ EXOR } p_{2,i,j,7}\right) \times 2^7 + \left(p_{1,\ i,j,6} \text{ EXOR } p_{2,i,j,6}\right) \times 2^6 + \cdots + \left(p_{1,i,j,0} \text{ EXOR } p_{2,i,j,0}\right) \times 2^0$$

When performed on binary images, EXORing gives the union of regions, less their intersection.

4. Local neighbourhood operators

Local neighbourhood operators are pixel mappings where each $q_{i,j}$ is derived from a region $\mathbf{N}_{i,j}$ in the neighbourhood of the corresponding $p_{i,j}$ in the input image:

$$q_{i,j} = f_{i,j}\left(\mathbf{N}_{i,j}\right)$$

Usually, the local area is centred around $p_{i,j}$, typically taking the form of a rectangular window of 3 × 3, 5 × 5 or 7 × 7 pixels.

Local neighbourhood operators are applied in identifying known shapes, reduction of noise and enhancement of edges.

(a) Linear filters

There is a well-established theory of linear filters, based on the mathematics of convolution (Dougherty and Giardina, 1987). Linear filters are an important class of neighbourhood operators, where each $p_{i,j}$ is replaced by a linear combination of its neighbours. Each of the neighbourhood pixels has a particular weighting by which its value is multiplied. The distribution of weights is known as a convolution mask or kernel. A 3 × 3 kernel is therefore:

$n_{-1,-1}$	$n_{-1,0}$	$n_{-1,1}$
$n_{0,-1}$	$n_{0,0}$	$n_{0,1}$
$n_{1,-1}$	$n_{1,0}$	$n_{1,1}$

where the central pixel $p_{i,j}$ is multiplied by weighting $n_{0,0}$; the pixel above and to the left of $p_{i,j}$ (that is $p_{i-1,j-1}$) is multiplied by $n_{-1,-1}$ and so on. The final output $q_{i,j}$ is obtained by summing the weighted neighbourhood values:

$$q_{i,j} = \sum_{k=-1}^{1} \sum_{l=-1}^{1} \left(p_{i+k,j+l} \times n_{k,l} \right)$$

The numerical value produced by the basic operation shown above is not guaranteed to lie in the interval [0, *W*]. Usually a scaling factor and offset are introduced. The following formulation restricts the output range to $0 \le q_{i,j} \le W$:

$$q_{i,j} = \frac{\sum_{k=-1}^{1} \sum_{l=-1}^{1} \left(p_{i+k,j+l} \times n_{k,l} \right)}{\sum_{k=-1}^{1} \sum_{l=-1}^{1} \left| n_{k,l} \right|} + \frac{W}{2} \left[1 - \frac{\sum_{k=-1}^{1} \sum_{l=-1}^{1} n_{k,l}}{\sum_{k=-1}^{1} \sum_{l=-1}^{1} \left| n_{k,l} \right|} \right]$$

If $p_{i,j}$ and $q_{i,j}$ are confined to the range

$$-\frac{W}{2} \le p_{i,j}, q_{i,j} \le \frac{W}{2}$$

the scaling operation is simplified to:

$$q_{i,j} = \frac{\sum_{k=-1}^{1} \sum_{l=-1}^{1} \left(p_{i+k,j+l} \times n_{k,l} \right)}{\sum_{k=-1}^{1} \sum_{l=-1}^{1} \left| n_{k,l} \right|}$$

Consider an image that consists of a single bright pixel, *p*, on a black background (Figure 11). Convolving this image with the kernel

1	1	1
1	1	1
1	1	1

results in the image shown in Figure 12.

0	0	0	0	0	0	0	0	0	0	0	0
0	0	0	0	0	0	0	0	0	0	0	0
0	0	0	0	0	0	0	0	0	0	0	0
0	0	0	0	0	0	0	0	0	0	0	0
0	0	0	0	0	0	0	0	0	0	0	0
0	0	0	0	0	*p*	0	0	0	0	0	0
0	0	0	0	0	0	0	0	0	0	0	0
0	0	0	0	0	0	0	0	0	0	0	0
0	0	0	0	0	0	0	0	0	0	0	0
0	0	0	0	0	0	0	0	0	0	0	0
0	0	0	0	0	0	0	0	0	0	0	0

Figure 11. Single pixel on black background.

In other words, the filter has caused the point to spread out to its neighbouring pixels. In this case, each pixel is replaced by its average value. Of course this is a highly simplified example to illustrate that in general each pixel in the output image will have a value that is contributed to, in part, by all of its eight neighbours. In this case where the

0	0	0	0	0	0	0	0	0	0	0	0
0	0	0	0	0	0	0	0	0	0	0	0
0	0	0	0	0	0	0	0	0	0	0	0
0	0	0	0	0	0	0	0	0	0	0	0
0	0	0	0	**p/9**	**p/9**	**p/9**	0	0	0	0	0
0	0	0	0	**p/9**	**p/9**	**p/9**	0	0	0	0	0
0	0	0	0	**p/9**	**p/9**	**p/9**	0	0	0	0	0
0	0	0	0	0	0	0	0	0	0	0	0
0	0	0	0	0	0	0	0	0	0	0	0
0	0	0	0	0	0	0	0	0	0	0	0
0	0	0	0	0	0	0	0	0	0	0	0

Figure 12. Effect of averaging filter.

output pixel is the average of its neighbourhood, the image is smoothed or blurred by this operation. In practice, the advantage of using this particular kernel is that any random noise in the image will be reduced.

A common variation is the Gaussian low-pass filter, of which an example of a 3 × 3 kernel is :

1	2	1
2	4	2
1	2	1

and is considered to have a smoother characteristic than the sharp cut-off of the averaging filter.

Now consider the following image of a region of uniform brightness, p (Figure 13). Convolving this image with the differencing kernel

−1	−1	−1
−1	8	−1
−1	−1	−1

results in the image of Figure 14 where $p_{i,j}$ and $q_{i,j}$ are confined to the range

$$-\frac{W}{2} \leq p_{i,j}, q_{i,j} \leq \frac{W}{2}$$

and $K = 16$.

Here, the boundary of the foreground block has been derived. The gradient of the edge changes sign from negative to positive as the kernel moves into the region. Where the kernel is completely inside the foreground, there is no difference in local grey level intensity, and the filter output is zero. The kernel shown above is known as a Laplacian filter. A simple modification of the Laplacian:

−1	−1	−1
−1	9	−1
−1	−1	−1

0	0	0	0	0	0	0	0	0	0	0	0
0	0	0	0	0	0	0	0	0	0	0	0
0	0	0	0	0	0	0	0	0	0	0	0
0	0	0	*p*	*p*	*p*	*p*	*p*	*p*	0	0	0
0	0	0	*p*	*p*	*p*	*p*	*p*	*p*	0	0	0
0	0	0	*p*	*p*	*p*	*p*	*p*	*p*	0	0	0
0	0	0	*p*	*p*	*p*	*p*	*p*	*p*	0	0	0
0	0	0	*p*	*p*	*p*	*p*	*p*	*p*	0	0	0
0	0	0	0	0	0	0	0	0	0	0	0
0	0	0	0	0	0	0	0	0	0	0	0
0	0	0	0	0	0	0	0	0	0	0	0

Figure 13. Block of uniform brightness.

0	0	0	0	0	0	0	0	0	0	0	0
0	0	0	0	0	0	0	0	0	0	0	0
0	0	-p/K	-2p/K	-3p/K	-3p/K	-3p/K	-3p/K	-2p/K	-p/K	0	0
0	0	-2p/K	5p/K	3p/K	3p/K	3p/K	3p/K	5p/K	-2p/K	0	0
0	0	-3p/K	3p/K	0	0	0	0	3p/K	-3p/K	0	0
0	0	-3p/K	3p/K	0	0	0	0	3p/K	-3p/K	0	0
0	0	-3p/K	3p/K	0	0	0	0	3p/K	-3p/K	0	0
0	0	-2p/K	5p/K	3p/K	3p/K	3p/K	3p/K	5p/K	-2p/K	0	0
0	0	-p/K	-2p/K	-3p/K	-3p/K	-3p/K	-3p/K	-2p/K	-p/K	0	0
0	0	0	0	0	0	0	0	0	0	0	0
0	0	0	0	0	0	0	0	0	0	0	0

Figure 14. Effect of high-pass filter.

gives a kernel that in effect adds the original image to the high-pass filtered version. This is a powerful edge enhancement filter, analogous to the photographic technique known as *unsharp masking*.

There is also a family of high-pass filters based on differences of Gaussian kernels. Edges are identified by the positions of zero-crossings in the filtered image.

Vertical and horizontal edges can be selectively enhanced by kernels such as the vertical edge detector and the horizontal edge detector:

1	0	−1
2	0	−2
1	0	−1

Vertical edge detector

1	2	1
0	0	0
−1	−2	−1

Horizontal edge detector

An example of an anisotropic linear filter is known as a radial filter, where the direction of orientation of the kernel elements depends on the spatial location over the space of the image. If a circular object is expected to be centralised in the image, the kernel is rotated such that the coefficient n2 is always orientated towards the centre of the image as shown in Figure 15. In this case, eight segments or octants are defined about the centre of the image, one for each possible orientation of the kernel. A more advanced modification of this filter might rotate the kernel about the centroid of an object in the field of view.

Linear filters have many useful applications, however trade-offs have to be made between the effectiveness of the filter and signal-to-noise ratio. When smoothing an image, the signal-to-noise ratio is improved but edges in the image lose definition. Conversely, linear filters intended to enhance edges suffer from sensitivity to noise.

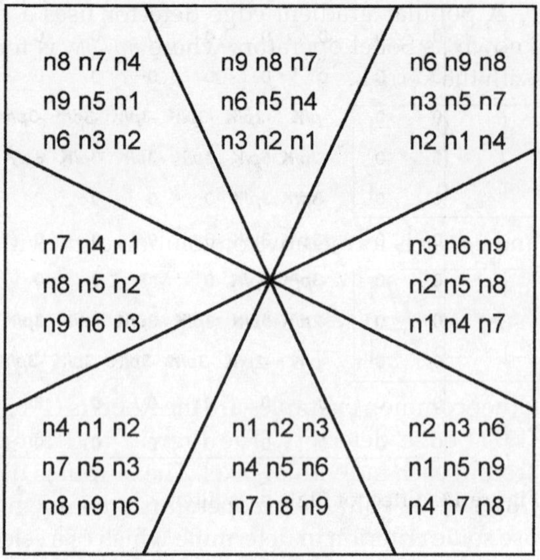

Figure 15. Radial filter coefficients.

(b) Non-linear filters

Many non-linear filters have been designed to preserve or enhance edge information while improving signal-to-noise ratio. In the simplest case, linear filters can be modified to omit *hot spots* (extremely bright or dark pixels) from the neighbourhood calculation, and are known as *trimmed filters.*

(i) Gradient Edge Detectors

Linear filters can also be used in non-linear combinations. Gradient edge detectors use a pair of linear filters, one to detect vertical edges and the other to detect horizontal. The pair of orthogonal derivatives are combined to give edge orientation and rate of change in the x and y directions.

If $\partial p/\partial x$ and $\partial p/\partial y$ are the rates of change in the x and y directions, respectively, the rate of change along a vector r, in the direction θ as measured from the x-axis is:

$$\frac{\partial p}{\partial r} = \frac{\partial p}{\partial x}\cos\theta + \frac{\partial p}{\partial y}\sin\theta$$

The direction in which the rate of change is greatest is:

$$\theta(x,y) = \tan^{-1}\left[\frac{\partial p/\partial y}{\partial p/\partial x}\right]$$

and the magnitude of the gradient in that direction is:

$$G[p(x,y)] = \sqrt{\left(\partial p/\partial x\right)^2 + \left(\partial p/\partial y\right)^2}$$

A popular gradient edge detector uses a pair of convolution kernels known as Sobel operators where $\partial p/\partial x$ is formed by convolution of $p_{i,j}$ with the kernel

1	0	−1
2	0	−2
1	0	−1

and $\partial p/\partial y$ is formed by convolution of $p_{i,j}$ with

1	2	1
0	0	0
−1	−2	−1

Other common examples are the Roberts (1965), Prewitt (1970) and Kirsch (1971) edge detectors. The Prewitt and Kirsch filters rotate the kernel around each individual pixel. The output is the absolute of the maximum convolution value. Such operators are known as *gated filters* because they use some criterion to determine which of a selection of linear filters will be used to produce the final output pixel value.

Prewitt filter coefficients:

1	1	1
1	−2	1
−1	−1	−1

Kirsch filter coefficients:

5	5	5
−3	0	−3
−3	−3	−3

(ii) Rank Filters

Rank filters are a useful class of non-linear filter, where the pixel values in a local neighbourhood are sorted according to value. The $p_{i,j}$ values inside a neighbourhood of **N** pixels are sorted into ascending numerical order:

$$(n_1, n_2, n_3, \cdots n_N)$$

where

$$(n_1 \leq n_2 \leq n_3 \leq \cdots n_N)$$

The output value R_k is selected by its maximum, median, minimum or any intermediate position in the ranked order:

$$R_k = n_k$$

Applying this filter to the entire image:

$$q_{i,j} = R_k\left(p_{i,j}\right)$$

Minimum, Median and Maximum Rank Filters

Consider a 3 × 3 neighbourhood of an image (Figure 16). The ranked order of pixel values is:

*	*	*	*	*	*	*
*	*	*	*	*	*	*
*	*	9	127	45	*	*
*	*	3	15	57	*	*
*	*	80	29	7	*	*
*	*	*	*	*	*	*
*	*	*	*	*	*	*

Figure 16. Local neighbourhood of image.

3, 7, 9, 15, 29, 45, 57, 80, 127

A filter output of rank 1 is known as a minimum filter (giving the value 3, in the above case). A ranking of **N**, where **N** is the number of pixels in the neighbourhood, corresponds to a maximum filter. A useful case is the median filter when **N** is odd, where the median rank is (**N**+1)/2. Median filtering is particularly useful in eliminating impulse spikes (single pixels with extreme values) without disrupting the shape of edges.

Consider the uniform image containing "salt-and-pepper" impulse noise (Figure 17).

p	*p*	*p*	*p*	*p*	*p*	*p*	*p*	*p*	*p*	*p*	*p*
p	*p*	*p*	*p*	*p*	*p*	*p*	*p*	*p*	*p*	*p*	*p*
p	*p*	**0**	*p*	*p*	*p*	*p*	*p*	***W***	*p*	*p*	*p*
p	*p*	*p*	*p*	*p*	*p*	*p*	*p*	*p*	**0**	*p*	*p*
p	*p*	*p*	*p*	*p*	**0**	*p*	*p*	*p*	*p*	*p*	*p*
p	***W***	*p*	*p*	**0**	*p*	*p*	***W***	*p*	*p*	*p*	*p*
p	*p*	*p*	*p*	*p*	*p*	*p*	*p*	*p*	**0**	*p*	*p*
p	*p*	*p*	*p*	*p*	*p*	*p*	*p*	*p*	*p*	*p*	*p*
p	*p*	*p*	*p*	*p*	***W***	*p*	*p*	*p*	*p*	*p*	*p*
p	*p*	*p*	*p*	*p*	*p*	*p*	*p*	*p*	*p*	*p*	*p*
p	*p*	*p*	*p*	*p*	*p*	*p*	*p*	*p*	*p*	*p*	*p*

Figure 17. Image containing impulse noise.

p	*p*	*p*	*p*	*p*	*p*	*p*	*p*	*p*	*p*	*p*	*p*
p	*p*	*p*	*p*	*p*	*p*	*p*	*W*	*W*	*W*	*p*	*p*
p	*p*	*p*	*p*	*p*	*p*	*p*	*W*	*W*	*W*	*p*	*p*
p	*p*	*p*	*p*	*p*	*p*	*p*	*W*	*W*	*W*	*p*	*p*
W	*W*	*W*	*p*	*p*	*p*	*W*	*W*	*W*	*p*	*p*	*p*
W	*W*	*W*	*p*	*p*	*p*	*W*	*W*	*W*	*p*	*p*	*p*
W	*W*	*W*	*p*	*p*	*p*	*W*	*W*	*W*	*p*	*p*	*p*
p	*p*	*p*	*p*	*W*	*W*	*W*	*p*	*p*	*p*	*p*	*p*
p	*p*	*p*	*p*	*W*	*W*	*W*	*p*	*p*	*p*	*p*	*p*
p	*p*	*p*	*p*	*W*	*W*	*W*	*p*	*p*	*p*	*p*	*p*
p	*p*	*p*	*p*	*p*	*p*	*p*	*p*	*p*	*p*	*p*	*p*

Figure 18. Maximum filtered image.

Maximum, minimum and median 3 × 3 window filters yield the results shown in Figures 18–20.

(c) Logical filters

Logical filters operate on binary images where each pixel assumes only two possible values, which for the purposes of analysis we shall take as 0

p	*p*	*p*	*p*	*p*	*p*	*p*	*p*	*p*	*p*	*p*	*p*
p	0	0	0	*p*	*p*	*p*	*p*	*p*	*p*	*p*	*p*
p	0	0	0	*p*	*p*	*p*	*p*	0	0	0	*p*
p	0	0	0	0	0	0	*p*	0	0	0	*p*
p	*p*	*p*	0	0	0	0	*p*	0	0	0	*p*
p	*p*	*p*	0	0	0	0	*p*	0	0	0	*p*
p	*p*	*p*	0	0	0	*p*	*p*	0	0	0	*p*
p	*p*	*p*	*p*	*p*	*p*	*p*	*p*	0	0	0	*p*
p	*p*	*p*	*p*	*p*	*p*	*p*	*p*	*p*	*p*	*p*	*p*
p	*p*	*p*	*p*	*p*	*p*	*p*	*p*	*p*	*p*	*p*	*p*
p	*p*	*p*	*p*	*p*	*p*	*p*	*p*	*p*	*p*	*p*	*p*

Figure 19. Minimum filtered image.

p	p	p	p	p	p	p	p	p	p	p	p
p	p	p	p	p	p	p	p	p	p	p	p
p	p	**p**	p	p	p	p	p	**p**	p	p	p
p	p	p	p	p	p	p	p	**p**	p	p	p
p	p	p	p	p	**p**	p	p	p	p	p	p
p	**p**	p	p	**p**	p	p	**p**	p	p	p	p
p	p	p	p	p	p	p	p	p	**p**	p	p
p	p	p	p	p	p	p	p	p	p	p	p
p	p	p	p	p	**p**	p	p	p	p	p	p
p	p	p	p	p	p	p	p	p	p	p	p
p	p	p	p	p	p	p	p	p	p	p	p

Figure 20. Median filtered image.

and 1. Binary images are typically produced towards the end of the image segmentation process. Logical filters can be used to "clean up" binary images contaminated with noise, and can extract or isolate features of interest (see Figure 21).

Consider a window of 3 × 3 pixels, centred around $p_{i,j}$,

$p_{i-1,j-1}$	$p_{i-1,j}$	$p_{i-1,j+1}$
$p_{i,j-1}$	$p_{i,j}$	$p_{i,j+1}$
$p_{i+1,j-1}$	$p_{i+1,j}$	$p_{i+1,j+1}$

The central pixel and its eight neighbours will form various binary patterns of 0 and 1, for example:

0	0	0
0	1	0
0	0	0

is an isolated point – possibly a result of noise in the original image.

Edges of objects might be represented by patterns such as:

1 1 0	0 1 1	1 0 0
1 1 0	0 1 1	1 1 1
1 0 0	1 1 1	1 1 1

The total number of possible combinations of bits in this 3 × 3 window is $2^9 = 512$ patterns. In practice, therefore, 512-entry look-up tables (LUTs) are often used to implement logical filtering operations. The central pixel and its eight neighbours are used to form a 9-bit pattern that directly addresses the LUT. Producing a 1-bit output, the LUT can store any

(a)

(b)

(c)

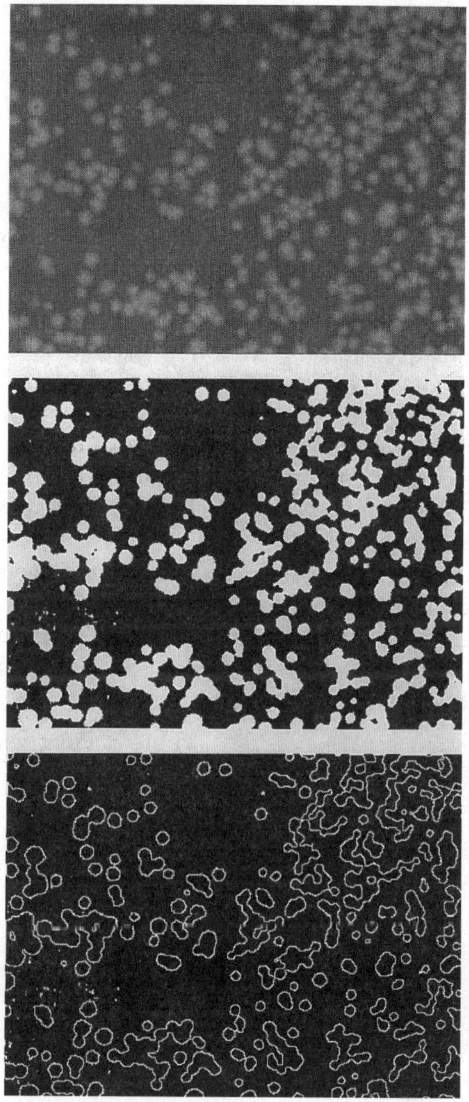

Figure 21. Effect of logical filters on colony image: (a) evenly illuminated image of *E. coli*; (b) threshold image and isolated point removal; (c) binary edge detection.

arbitrary filter. When the address corresponds to a neighbourhood pattern of interest, the LUT entry is preset to logical 1, otherwise logical 0.

In practical applications, real-time processing is possible because the image is produced by the camera and A/D converter as a serial train of pixel values. Tapped shift registers are used to access the 3 × 3 neighbourhood and present it to the LUT. The output of the LUT is immediately available at video rates.

We shall now consider the design of logical filters.

For convenience, we shall represent the bit pattern corresponding to the local neighbourhood of $p_{i,j}$ with the notation:

m1	m2	m3
m4	m5	m6
m7	m8	m9

where m5 = $p_{i,j}$, m1 = $p_{i-1,\,j-1}$, etc.

(i) Point Remove

If the central pixel is set, and all of the surrounding pixels are not set, then the central pixel is an isolated point:

m1	m2	m3		0	0	0
m4	m5	m6	=	0	1	0
m7	m8	m9		0	0	0

To remove isolated points we must detect the above combination and set the corresponding LUT value to logical 0. Any other combinations where m5 is set must be left unchanged – these will represent the edges or the interiors of objects.

Rather than forming the complete 512 element table, for the purpose of example we shall consider only the neighbours m4, m5 and m6, and we shall define the combination 0 1 0 as an isolated point. In this case we have an 8-element look-up table. To delete only isolated points, the entries shown in Table 1 are required.

Table 1. Point remove

$p_{i,j}$	m4	m5	m6
0	0	0	0
0	0	0	1
0	0	1	0
1	0	1	1
0	1	0	0
0	1	0	1
1	1	1	0
1	1	1	1

It is possible to form a logical expression describing the required operation, using the rules of Boolean algebra, where "+" represents "OR", "• " represents "AND", and "$\bar{\text{m}}$" represents "NOT m". By inspection of the table, the conditions that describe $q_{i,j}$ are:

$$q_{i,j} = (\overline{\text{m4}} \bullet \text{m5} \bullet \text{m6}) + (\text{m4} \bullet \text{m5} \bullet \overline{\text{m6}}) + (\text{m4} \bullet \text{m5} \bullet \text{m6})$$
$$q_{i,j} = \text{m5} \bullet (\text{m4} \bullet \overline{\text{m6}} + \overline{\text{m4}} \bullet \text{m6} + \text{m4} \bullet \text{m6})$$
$$q_{i,j} = \text{m5} \bullet \left(\text{m4} \bullet \overline{\text{m6}} \bullet \text{m6} \bullet \left(\overline{\text{m4}} + \text{m4}\right)\right)$$
$$q_{i,j} = \text{m5} \bullet (\text{m4} \bullet \overline{\text{m6}} + \text{m6})$$
$$q_{i,j} = \text{m5} \bullet (\text{m4} + \text{m6})$$

Expanding the above example to include the full 3 × 3 window, the expression becomes:

$$q_{i,j} = \text{m5} \bullet (\text{m1} + \text{m2} + \text{m3} + \text{m4} + \text{m6} + \text{m7} + \text{m8} + \text{m9})$$

If m5 and any of its neighbours are set, $q_{i,j}$ is assigned a value of logical 1. If m5 alone is set, $q_{i,j}$ is zeroed and the isolated point is deleted.

(ii) Binary Edge Detect

If the central pixel is set, and any one of its neighbours is not set, then the central pixel is an edge, as illustrated earlier. In our simplified look-up table, the entries corresponding to edges are itemised in Table 2.

Table 2. Binary edge detect

$p_{i,j}$	m4	m5	m6
0	0	0	0
0	0	0	1
1	0	1	0
1	0	1	1
0	1	0	0
0	1	0	1
1	1	1	0
0	1	1	1

$$q_{i,j} = (\overline{m4} \bullet m5 \bullet \overline{m6}) + (\overline{m4} \bullet m5 \bullet m6) + (m4 \bullet m5 \bullet \overline{m6})$$
$$q_{i,j} = m5 \bullet (\overline{m4} \bullet (\overline{m6} + m6) + m4 \bullet \overline{m6})$$
$$q_{i,j} = m5 \bullet (\overline{m4} + m4 \bullet \overline{m6})$$
$$q_{i,j} = m5 \bullet (\overline{m4 \bullet m6})$$

Expanding to cover the full 3 × 3 window:

$$q_{i,j} = m5 \bullet \overline{(m1 \bullet m2 \bullet m3 \bullet m4 \bullet m6 \bullet m7 \bullet m8 \bullet m9)}$$

If m5 is set, then $q_{i,j}$ is set, with the exception where all of m5's neighbours are simultaneously set to logical 1.

(iii) Erosion

Regions in a binary image can be reduced in area by a process known as erosion, where pixels at the boundary of the object are deleted. The simplified look-up table is shown in Table 3.

Table 3. Erosion

$p_{i,j}$	m4	m5	m6
0	0	0	0
0	0	0	1
0	0	1	0
0	0	1	1
0	1	0	0
0	1	0	1
0	1	1	0
1	1	1	1

$q_{i,j}$ = m4 • m5 • m6

Expanding the above operation to cover the full 3 × 3 window:

$q_{i,j}$ = m1 • m2 • m3 • m4 • m5 • m6 • m7 • m8 • m9

$q_{i,j}$ is set only when m5 and all of its neighbours are set.

(iv) Dilation

Dilation is the inverse operation to erosion, increasing the area of a region:

Table 4. Dilation

$p_{i,j}$	m4	m5	m6
0	0	0	0
1	0	0	1
1	0	1	0
1	0	1	1
1	1	0	0
1	1	0	1
1	1	1	0
1	1	1	1

$q_{i,j}$ = m4 + m5 + m6

Expanding to cover the full 3 × 3 window:

$q_{i,j}$ = m1 + m2 + m3 + m4 + m5 + m6 + m7 + m8 + m9

Holes in a region can be eliminated by one or more iterations of dilation, which fills the hole, followed by the same number of iterations of erosion, which restores the overall boundary of the region to its original size.

(v) Connectivity

By definition, a region is a collection of pixels that are connected together. By working through the neighbouring pixels, it is possible fully to explore the entire region. Regions are described as either 4-connected, or 8-connected depending on whether the four immediate neighbours corresponding to the directions North, South, East and West are considered, or whether all eight immediate neighbours are taken.

The following patterns are 4-connected:

0	1	0		0	0	0		0	1	0
0	1	0		1	1	1		0	1	1
0	1	0		0	1	0		0	0	0

The following are 8-connected:

1	0	1		0	0	1		1	1	0
0	1	0		1	1	1		0	1	1
1	0	0		0	0	1		0	0	0

(vi) Critical Connectivity

In the above six patterns, there are instances where if the central pixel were reset to logical 0, the connected region would be split apart.

Consider the 8-connected examples where the central pixel has been reset:

1	0	1		0	0	1		1	1	0
0	0	0		1	0	1		0	0	1
1	0	0		0	0	1		0	0	0

In the first case, three separate regions are produced, in the second case there are two regions, and only in the final case has 8-connectivity been preserved.

Therefore patterns such as:

1	0	1		0	0	1
0	X	0	and	1	X	1
1	0	0		0	0	1

are regarded as critical for 8-connectivity, while patterns such as:

1	1	1		0	0	1
1	X	0	and	1	X	1
1	0	0		0	0	1

are not critical.

Look-up tables can be designed that incorporate patterns critical for connectivity.

(d) Morphological image processing

Erosion and dilation as shown above are special cases of morphological image processing operators. Morphological operators use a template known as a *structuring element* that is applied to each pixel, $p_{i,j}$. The template contains a pattern of 0's and 1's that are combined in an arbitrary logical function with the pixels in the corresponding positions in the original binary image.

Thus in the cases described above, the structuring element is a 3 × 3 region with the pattern:

1	1	1
1	1	1
1	1	1

that is, all of the pixels in this structuring element are involved in the calculation. The logical functions to be applied are:

- Isolated point removal:
  $q_{i,j} = m5 \bullet (m1 + m2 + m3 + m4 + m6 + m7 + m8 + m9)$
- Binary edge detect:
  $q_{i,j} = m5 \bullet \overline{(m1 \bullet m2 \bullet m3 \bullet m4 \bullet m6 \bullet m7 \bullet m8 \bullet m9)}$
- Erosion:
  $q_{i,j} = m1 \bullet m2 \bullet m3 \bullet m4 \bullet m5 \bullet m6 \bullet m7 \bullet m8 \bullet m9$
- Dilation:
  $q_{i,j} = m1 + m2 + m3 + m4 + m5 + m6 + m7 + m8 + m9$

In general, morphological operators use a structuring element of any size and shape. Features of the image that are expected to be present can be extracted by designing appropriate structuring elements. Morphological operations have their origins in the mathematics of set theory. There is an established *image algebra* in which binary and grey scale images are the operands (Dougherty and Giardina, 1987).

(e) Region labelling

Having obtained a "clean" binary image containing several connected regions, the remainder of the task is to analyse the components before taking appropriate action. At this stage it is useful to assign an identity or label to each of the separate regions. Region labelling is the process of assigning a unique grey level value to each object. The result of the operation is a grey scale image or *map* where each connected region has a particular label (Figure 22). The fourth object encountered will be assigned grey level value 4, for example. If required, the *n*th object can be separated out by binary thresholding.

In the Flexys™ Picker, the region labelling algorithm scans down through the image line by line from left to right. When an unlabelled region is encountered, the algorithm propagates the label value by exploring the region via the neighbouring pixels. When all the pixels in the region have been labelled, the scan resumes from where it left off, until a new unlabelled region is found. The label is then incremented in value and propagated as before.

B. Intermediate Level Image Processing

Intermediate level processing takes an iconic image, such as a region labelled map and abstracts from this a description of the properties of its components. This is a data reduction technique where the pictorial information is reduced to a set of descriptors describing the image features. The descriptors may be presented in the form of a data list. For example, a feature vector can be defined as an ordered n-tuple {x1, x2, x3, . . . xn} of scalar properties.

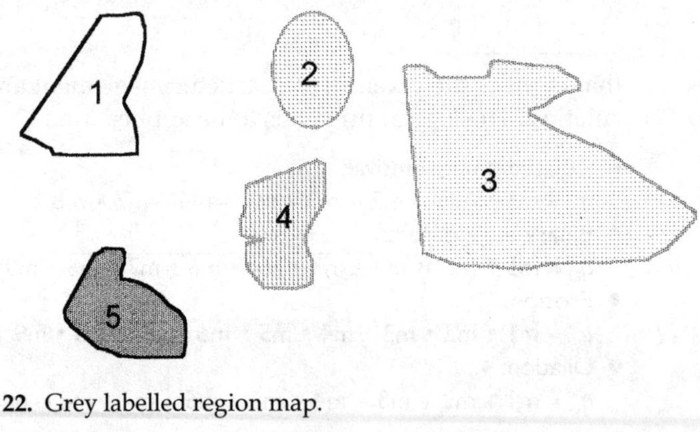

Figure 22. Grey labelled region map.

1. Property descriptors

A set of suitable shape descriptors has been developed at SRI International (Agin and Duda, 1975). These include:

1. Object Perimeter
2. Square Root of Area
3. Total Hole Area
4. Minimum Radius
5. Maximum Radius
6. Average Radius
7. Compactness Ratio (perimeter/square root of area)

Descriptors which are independent of scale, translation or rotation are particularly useful. These include the Euler number (number or regions less the number of holes), compactness ratio, number of holes, aspect ratio of minimum bounding rectangle, convex discrepancy and the invariant moments (Batchelor *et al.*, 1985; Gonzalez and Wintz, 1987).

(a) Invariant moments

An important set of useful shape descriptors are known as the invariant moments. If $p(i, j)$ is a digital image, the moment of order $(x + y)$ is:

$$M_{xy} = \sum_i \sum_j i^x j^y p(i,j)$$

For binary images, $p(i, j)$ takes only the values 0 and 1. The sum of all set pixels in the image is given by:

$$M_{00} = \sum_i \sum_j p(i,j)$$

and is usually interpreted as an area if $p(i, j)$ contains only one connected region.

(i) Centroid

The centroid of a region depends on its position in the image, given by the co-ordinates:

$$i' = M_{10} / M_{00}$$

$$j' = M_{01} / M_{00}$$

By centralising the computation around the centroid, the central moments μ_{xy} are independent of position:

$$\mu_{xy} = \sum_i \sum_j (i - i')^x (j - j)^y p(i,j)$$

It is possible to make the central moments independent of translation, rotation and scaling. As x and y assume various non-negative integer values, the set of moments so generated specify uniquely the shape of an object in the image.

(b) Polar distance

The polar distance is a radial measure of the distance from the centroid of an object to its boundary edge (Figure 23). As the angle of measurement varies from 0 to 360 degrees, a graph or signature is produced that describes the shape of the region. A perfect circle has a unique signature consisting of a straight line.

C. High-level Processing

At the high levels of processing, the descriptive symbols extracted earlier are interpreted, and appropriate action may be initiated. In many industrial machine vision applications, extensive computation at this level is not required. Typically, an automated visual inspection system simply verifies that the feature descriptors lie within specified tolerance limits. If they do not, a reject solenoid is activated. For object recognition, an extracted feature vector may be matched against the corresponding member of a stored database. Statistical pattern classification methods are often used to establish the degree of fit (Tou and Gonzalez, 1974).

The sequence of transformations from the original image to segmented structures, to symbolic representations and scene descriptions is often referred to as data driven or bottom-up processing. Emphasis is placed upon information extraction at the pixel level, which is automatic and independent of the higher stages. Bottom-up processing is reliable if the input data is noise free and repeatable. Intensive computation is required.

Model driven processing places the emphasis on knowledge of the task domain. Pixel-level operations are explicitly directed from the high level. Approaches such as this are known as goal directed or top-down processing. Typically, a knowledge base is used to predict structures expected in the original image. The hypotheses are subsequently verified, often using selective perception to concentrate on areas of particular interest. Allocation of system resources is therefore more efficient than with bottom-up control. Top-down processing is also more tolerant of noisy and unreliable input images. However, the technique is highly domain dependent and can misinterpret the data if the scene is outside the scope of the model.

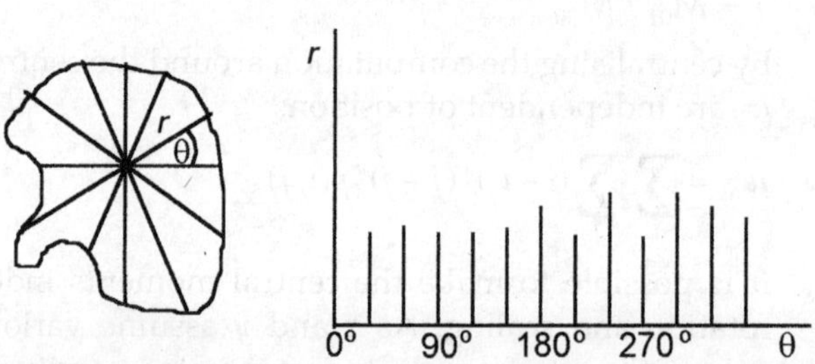

Figure 23. Polar distance.

◆◆◆◆◆◆ V. FLEXYS™ IMAGE PROCESSING ALGORITHM

The Flexys™ Colony and Plaque Picker scans images from the surface of various source plates and processes this data by looking at each individual pixel in the image. Controlled using adjustable image processing parameters, the analysis determines which objects in the field of view are to be picked and which are not. This section describes the operation of the set of image processing operators and how they can be used to select or deselect any given colony type.

The following operations are applied to the image in turn. All those objects that have shape descriptors falling inside the user-set limits are considered to be valid colonies which are then targeted to be picked.

A. Smoothing Window

The smoothing window is a linear averaging filter for reducing the background noise in the image. The smoothing kernel is:

n1	n2	n3		1	1	1
n4	n5	n6	=	1	1	1
n7	n8	n9		1	1	1

This will remove occasional noisy pixels from the image which might otherwise distort the shape of colonies. Although useful for reducing background noise, smoothing also has the effect of reducing edge definition by blurring the image. Too much smoothing will blur the image excessively, making the colonies difficult to detect. The smoothing kernel can be varied in size if required.

B. Local Threshold Difference (LTD)

The LTD detects boundaries of objects within the field of view, and is effectively a sensitivity setting. Colony or plaque boundaries are detected as the camera scans across the image looking for sharp changes in grey scale. The grey scale value of a colony is compared to the grey scale of the background (agar). If this difference is greater than the LTD value then a colony boundary is detected.

The LTD is implemented with a modified non-linear ranked filter $LTR(p_{i,j})$ where the output value is selected by subtracting the pixel of rank **N** from that of rank 1, where **N** is the number of elements in the neighbourhood. The minimum value in the neighbourhood is therefore subtracted from the maximum:

$$q_{i,j} = LTR\left(p_{i,j}\right) = R_N\left(p_{i,j}\right) - R_1\left(p_{i,j}\right)$$

Consider the 3 × 3 neighbourhood in the image shown in Figure 24. The ranked order of pixel values is:

3, 7, 9, 15, 29, 45, 57, 80, 127

*	*	*	*	*	*	*
*	*	*	*	*	*	*
*	*	9	127	45	*	*
*	*	3	15	57	*	*
*	*	80	29	7	*	*
*	*	*	*	*	*	*
*	*	*	*	*	*	*

Figure 24. Neighbourhood of image.

and the LTD filter output in this case is 127–3 = 124. The difference is then compared to a threshold, C, supplied by the user. If the difference is greater than C, the central pixel is marked as an edge point:

$$q_{i,j} = \begin{cases} 0, & \text{if } 0 \le \mathrm{LTR}(p_{i,j}) < C \\ W, & \text{if } C \le \mathrm{LTR}(p_{i,j}) \le W \end{cases}$$

The threshold value, C, is set to be more or less sensitive to differences in grey scale. A low value detects small differences in grey scale corresponding to low contrast colonies. High values will only detect the strongest differences in intensity, so ignoring less well defined colonies.

In the practical implementation of the local threshold difference operator, the neighbourhood window size as well as the difference threshold can be selected by the user.

The output of this operation is a binary image delineating the borders of objects with strong edges, as determined by the threshold value, C. For circular colonies and plaques, the border will be circular and the LTD will therefore be an annulus (Figure 25).

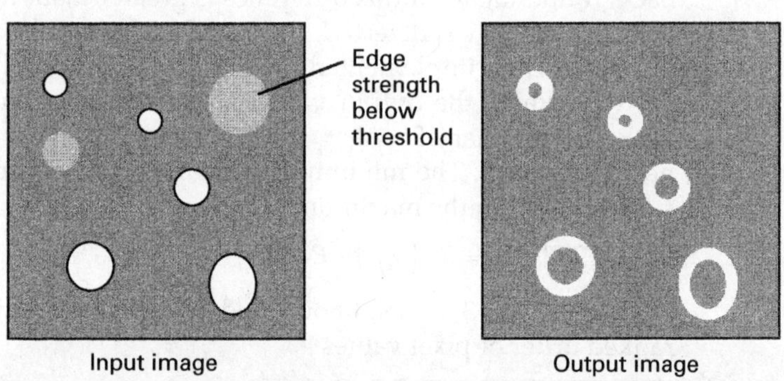

Figure 25. Effect of LTD operation.

C. Minimum and Maximum Area

Minimum and maximum area values relate to the size of objects found in the field of view, measured in pixels. Strictly, this is the area of the annulus generated by the LTD operation, rather than the total area of the colony.

D. Maximum Non-circularity

The maximum non-circularity parameter considers the shape of the colonies found in the image. This measures how far the colony deviates from a perfect circle. This is done by measuring the polar distance of pixels from the centre of the colony to the boundary in eight directions. Colonies are rejected if the deviation in the eight values exceeds a difference threshold (Figure 26).

Adjusting the threshold value determines how non-circular the colonies are permitted to be. Typically a value of 3 will reject all but the most uniform of *E. coli* colonies. Higher values can be set to tolerate grossly non-circular colonies such as some yeast colonies. High values should, however, be used with caution. Two touching or merged colonies may be seen as a figure of eight and can be tolerated by a high non-circularity value.

E. Maximum Grey Level and Minimum Grey Level

The maximum and minimum grey level parameters are the last to be applied to the scanned image. The grey level of the centre of a colony is compared to the average background grey level immediately surrounding it. The background intensity is subtracted from the colony level to give a positive or negative value. A white colony therefore tends to have a positive value while a blue colony has a negative value. The parameters can be set to exclude either blue or white colonies or any shade between. For example, a maximum of 255 and minimum of 20 will select only the white colonies to be picked.

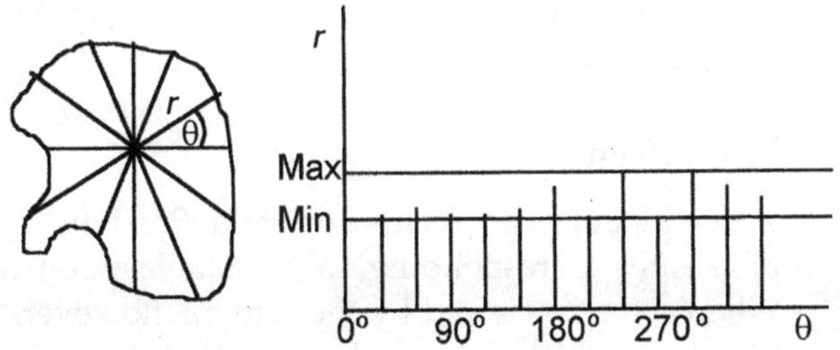

Figure 26. Circularity measure.

◆◆◆◆◆◆ VI. CO-ORDINATE CONVERSION

Throughout a picking run, the picking needles will be targeted on the centroid of each of the identified colonies or plaques. However, the co-ordinate system used by the image processing algorithm is based on the dimensions of the frame grabber (0–767 pixels horizontally, 0–575 vertically). These local co-ordinates in pixels must be converted to the corresponding real-world locations on the bed of the robot. A conversion algorithm is used to transform the frame-grabber co-ordinates into the physical dimensions in micrometres. During factory commissioning of the robot, an engineer enters a set of calibration positions and fixed offsets defining the locations of each of the plates, fixtures and tools. The fixed offsets are added to the converted frame-grabber co-ordinates to position the camera and picking tool at the correct real-world locations.

During the robot's working life, small deviations in the camera lens position and magnification can cause errors in co-ordinate conversion large enough to make the picking needles miss their targets. Also, from time to time the picking needles may be replaced or might be accidentally knocked slightly out of position. Fixed offsets with ideal values are therefore not sufficient to calibrate the robot in practical applications.

The Flexys™ therefore carries out a set of self-calibration procedures in which various patterns are punched by the picking tool, on to white paper mounted in a special frame. The image processing system calculates the positions of the punched holes and compares them with their expected ideal locations. The differences are used to calculate the translational, rotational and scaling errors introduced by the camera and lens. A separate tool calibration procedure compensates for any deviation of the picking needles. Before starting a picking run it is advisable for the operator to go through the camera and tool calibration process.

A. Camera Calibration

To compensate for errors at the camera, three holes are punched, corresponding to the vertices of a scalene triangle (Figure 27). Ideally, the vertices should be punched at co-ordinates:

$X_1 = 0\ \mu m$
$Y_1 = 0\ \mu m$
$X_2 = 18\,750\ \mu m$
$Y_2 = 28\,750\ \mu m$
$X_3 = 23\,750\ \mu m$
$Y_3 = 2500\ \mu m$

where X_1 and Y_1 is a relative origin defined during factory commissioning, physically corresponding to a suitable point on the calibration sheet. When actually viewed by the camera, however, the imaged co-ordinates may be translated, rotated and scaled, for example X_2 and Y_2 might be shifted to x_2 and y_2 as shown in Figure 27.

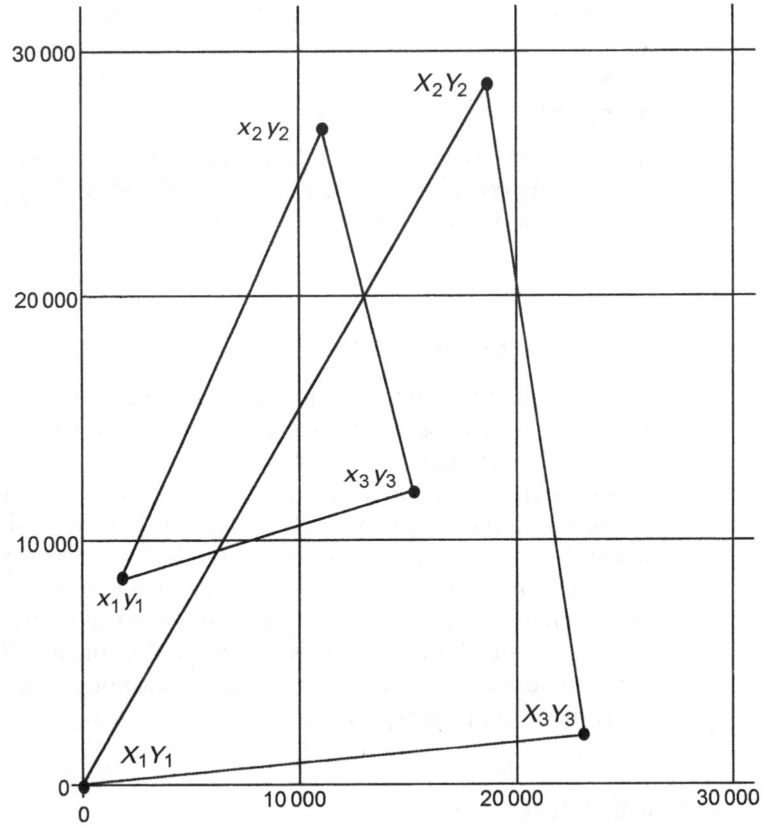

Figure 27. Camera calibration points.

To convert an arbitrary point x, y, as imaged by the camera, into the calibrated robot co-ordinates relative to X_1 and Y_1, the following transformation is created:

$$\begin{bmatrix} x_{\text{robot}} & y_{\text{robot}} \end{bmatrix} = \begin{bmatrix} x, y \end{bmatrix} \begin{bmatrix} S_{0,0} & S_{0,1} \\ S_{1,0} & S_{1,1} \end{bmatrix}$$

where

$$S_{0,0} = \frac{X_2/y'_2 - X_3/y'_3}{x'_2/y'_2 - x'_3/y'_3}$$

$$S_{0,1} = \frac{Y_2/y'_2 - Y_3/y'_3}{x'_2/y'_2 - x'_3/y'_3}$$

$$S_{1,0} = \frac{X_2/x'_2 - X_3/x'_3}{y'_2/x'_2 - y'_3/x'_3}$$

$$S_{1,1} = \frac{Y_2/x'_2 - Y_3/x'_3}{y'_2/x'_2 - y'_3/x'_3}$$

and where

$$x'_2 = x_2 - x_1$$
$$y'_2 = y_2 - y_1$$
$$x'_3 = x_3 - x_1$$
$$y'_3 = y_3 - y_1$$

are the vertices of the triangle imaged during the calibration process. Finally, the relative co-ordinates are translated to the global origin point of the robot by adding a pair of fixed offsets, defined during the commissioning process.

I. Advanced camera calibration

A second stage of camera calibration can be optionally carried out by the user, to compensate for any skew imposed as the camera is translated. In this case a rectangular 6 × 4 array of holes is punched over the area of a fresh calibration paper. This array acts as a series of control points, corresponding to the nodes of a grid. If the mechanical or imaging system is skewed in any way, the locations of the holes as seen by the camera will deviate from their ideal positions. Each hole will be subject to a local offset vector that is calculated during the advanced camera calibration procedure. A set of 24 local offsets is generated. During a picking run, the offsets are added piecewise to the calibrated colony co-ordinates from the corresponding local regions.

B. Tool Calibration

To calibrate the six-pin picking tool, a rectangular array of 36 holes is punched. Six holes are made per needle in the format shown in Figure 28. For each pin, the difference in the ideal and actual punched positions is calculated. An average of six offsets is taken per pin to compensate for any statistical spread in the targeting of the driving solenoids. During operation, each needle therefore has associated with it an average offset that is combined with the calibrated co-ordinates generated earlier by the imaging system.

1	2	3	4	5	6
6	1	2	3	4	5
5	6	1	2	3	4
4	5	6	1	2	3
3	4	5	6	1	2
2	3	4	5	6	1

Figure 28. Tool calibration pattern.

The 24-pin tool is calibrated similarly, except that the calibration pattern is an array of 6 × 4 holes. Each needle is activated 12 times without changing location. The centroid of the hole generated is therefore an average of the physical deviations in the punch position.

Acknowledgements

The authors would like to acknowledge the Flexys™ Picker development team, Robert M. Davies, Michael Stewart, John Evans, Andrew Watson, Gavin McKeown, Kanchi Karuntaratne, Martin Oliver, David Byatt and Brian Munday.

References

Agin, G. J. and Duda, R. O. (1975). SRI vision research for advanced industrial automation. *Proc. 2nd USA/Japan Comput. Conf.*, pp. 113–117.

Bassmann, H. and Besslich, P. W. (1995). *Ad Oculos Digital Image Processing Student Version 2.0*. International Thomson Publishing.

Batchelor, B. G. and Whelan, P. F. (1997). *Intelligent Vision Systems for Industry*. Springer-Verlag.

Batchelor, B. G., Hill, D. A. and Hodgson, D. C. (eds) (1985). *Automated Visual Inspection*. IFS (Publications) Ltd, Bedford, UK and North-Holland, Amsterdam.

Castleman, K. R. (1996). *Digital Image Processing*, 2nd edn. Prentice-Hall, New Jersey.

Dougherty, E. R. and Giardina, C. R. (1987). *Matrix Structured Image Processing*. Prentice-Hall, New Jersey.

Gonzalez, R. C. and Wintz, P. (1987). *Digital Image Processing*, 2nd edn. Addison-Wesley, Massachusetts.

Kirsch, R. (1971). Computer determination of the constituent structure of biological images. *Comp. Biomed. Res.*, **4**, 315–328.

Pratt, W. K. (1991). *Digital Image Processing*, 2nd edn. John Wiley, New York.

Prewitt, J. M. S. (1970). Object enhancement and extraction. In *Picture Processing and Psychopictorics* (B. S. Lipkin and A. Rosenfeld, eds.), pp. 75–149. Academic Press, New York.

Roberts, L. G. (1965). Machine perception of three-dimensional solids. In *Optical and Electro-Optical Information Processing* (J. T. Tippett, ed.), pp. 159–197. MIT Press.

Rosenfeld, A. and Kak, A. C. (1982). *Digital Picture Processing*, 2nd edn., Vols 1 and 2. Academic Press, New York.

Stewart, M., Watson, A., McKeown, G., Karuntaratne, K., Evans, J., Oliver, M. and Davies, R. M. (1995). A general purpose instrument control system and its application to a flexible laboratory robot, *LRA* **7**, 85–91.

Tou, J. T. and Gonzalez, R. C. (1974). *Pattern Recognition Principles*. Addison-Wesley, Massachusetts.

3 Library Picking, Presentation and Analysis

David R. Bancroft[1], Elmar Maier[1] and Hans Lehrach[2]

[1] GPC AG Genome Pharmaceuticals Corporation, Lockhamer Straße 29, D-82152 Martinsreid, Germany [2] Max-Planck-Institut für Molekulare Genetik, Ihnestraße 73, Berlin-Dahlem, Germany

Library Picking

CONTENTS

I. STATISTICS, SCALE AND STRATEGY

A. Statistical Considerations

The genomes of higher organisms contain an immense amount of information – the human genome consists of 3×10^9 bp. Only a very small subset of this information is transcribed and translated, resulting in highly different levels of developmentally-, tissue- and pathologically-specific expression patterns.

When genetic libraries are constructed, genomic DNA or transcribed mRNA is extracted from source tissues and ultimately cloned into a suitable genetic system. This process can be considered a random sampling event, where a small number of nucleic acid fragments are sampled from a large and heterogeneous population. Such a treatment allows the scientist to estimate the statistical characteristics of a genetic library.

B. Overall Library Size

Using simple sampling models, the probability of cloning a particular fragment of genomic DNA can be predicted as a function of three parameters: the total genome size of the study organism; the average insert size for a given cloning system; and the total number of clones in the library. The shape of this function is shown in Figure 1(a) and clearly displays two factors: (i) for genome sizes typically found in higher eukaryotes, hundreds of thousands of clones need to be analysed; and (ii) the function curves asymptotically towards 100%, meaning that by investigating higher numbers of clones, only diminishing returns are obtained.

Similar models can be applied to predict the size of primary cDNA libraries, where the probability of cloning a particular transcript can be estimated from two factors: (i) the relative abundance of the transcript in the mRNA pool and (ii) the total library size. Figure 1(b) displays this function for a typical range of transcript abundances, and indicates that libraries approaching 500 000 clones are required in order to obtain a reasonable likelihood of cloning a rare transcript at least once.

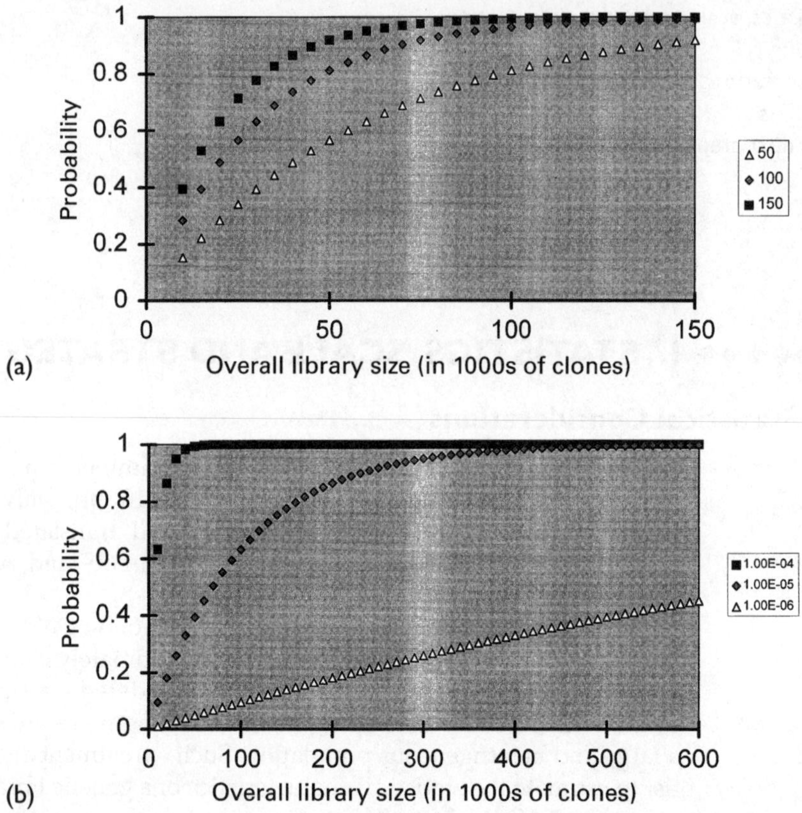

Figure 1. (a) The probability of cloning a given DNA fragment as a function of total haploid genome size G, the average insert size I and the overall size of the genomic library N. The function is described by $P = 1 - (1 - (I/G))^N$ (derived from Glover, 1984) and applied here using a genome size of 3×10^9 bp, and average insert sizes typical of cosmid (50 kb), P1 (100 kb) and PAC/BAC (150 kb) cloning systems (see key) (b) The probability that a mRNA will be found in a cDNA library as a function of the overall size of the cDNA library N and the fractional proportion (see key) of mRNA molecules of interest f. The function is described by $P = 1 - (1 - f)^N$ (derived from Sambrook *et al.*, 1989).

It is apparent from such statistical treatment that huge numbers of clones must be investigated in the course of a typical genome investigation. Yet, genetic libraries of several million clones are routinely handled in laboratories without any automation whatsoever. However, these libraries are constructed and treated as a single, mixed and unordered sample of clones. Handling this number of clones becomes a significant logistical problem only if each clone is to be treated as a unique entity. The scale of this logistical problem can only be overcome by large-scale automated processes and the use of arrayed libraries

C. Arrayed Libraries and High-throughput Strategies

Over recent years the use of arrayed clone libraries, where each clone is stored as an individual entity in a well of a microtitre plate which can be subsequently retrieved, has become an established tool in most genome analysis laboratories (Gress *et al.*, 1992; Hoheisel *et al.*, 1993; Lennon and Lehrach, 1991). On the whole, the use of arrayed libraries has two main advantages. First, each clone that is analysed, whether interesting in the context of a particular experiment or not, exists as a permanent reference in form of a frozen stock, meaning that usually no secondary rounds of screening are required, once an address in a microwell plate has been identified. Second, an infinite number of copies of a library can be made, so that the same biological resource can be distributed to and shared among other investigators. This can have the huge advantage that data generated in many laboratories in the world using quite different experimental strategies can be linked via a common factor, namely the clone library and associated clone addresses. Based upon this principle, several centres have been established that specialise in the distribution of arrayed libraries (Lennon *et al.*, 1996; Zehetner and Lehrach, 1994). The generation and analysis of large arrayed libraries requires a high level of automation (Maier *et al.*, 1994b), which is described later in the chapter.

In the fields of automation and production, there are two general approaches in designing high-throughput systems:

1. Serial processes capable of dealing with a small number of clones in a single experiment, but one in which the component steps can be easily streamlined. An example of such a process being conventional gel based sequencing (and, increasingly, capillary electrophoretic systems) where a few tens of clones are sequenced in single gel run.
2. Highly parallel processes in which tens of thousands of samples are processed and analysed in a single experiment. An example of such a strategy from genome research is the use of high-density hybridisation filters, when tens or hundreds of thousands of individual genetic samples can be screened in a single hybridisation experiment.

In this chapter we will describe the automated systems developed, and biological considerations required to implement such a highly parallel screening process based on hybridisation of high-density arrays.

◆◆◆◆◆◆ II. PICKING

A. Robotic Hardware

The picking of individual clones makes an important conceptual leap: the transition from the random arrangements, mixtures or collections of clones within a raw transformation, into an ordered array of individual clones within microtitre plates suitable for long-term storage, analysis and individual retrieval. In recent years several clone picking systems have been designed (Uber *et al.*, 1991; Jones *et al.*, 1992) which use a single or a 4-pin picking head. Solutions other than pin-based picking have been proposed, proved and in some cases successfully applied. For example, the use of disposable plastic tubing to pick phage plaques (Mardis *et al.*, 1995), cell sorting using modified FACS technology (van den Engh *et al.*, 1995) and individual cell attachment using microstructured binding supports (O'Neil *et al.*, 1997). However, pin picking has proved the most widely used approach and forms the basis of the picking system described here.

Our clone picking robot differs in several crucial aspects from these developments. We have integrated the clone picking feature in our flatbed robot system of the gridding robot which will be described later. The picking is conducted using a picking manifold with 96 spring loaded pins arranged to fit into a microtitre plate, where each pin can be individually extended into a colony using a pneumatic cylinder. Since plate inoculation and pin sterilisation are the two steps most limiting to the picking rate, our system is several times faster than other devices. The system is capable of picking approximately 3500 *E. coli* clones per hour into 384 microtitre plates. Two large colony trays (225 mm × 225 mm) and twenty-four 384-well microtitre plates can be positioned on the robot bed (Figure 2).

B. Vision Software

Automated picking is only possible by intelligent and fast computer imaging systems. A CCD camera affixed to the robotic head captures an image of a region of the colony tray. This image is rapidly processed by the imaging software to identify colonies, the locations of which are fed back to the motion system to enable picking. Three important characteristics of a vision system must be combined in order to maximise the accuracy and efficiency of picking:

1. Calibration systems to relate the relative pixel positions of a computer image into real-world positions on the bed of the robot. In our systems this takes the form of separate calibration functions to account for image distortion from lens barrelling, functions to relate the camera position to the picking position of individual pins and, finally, calibration of the distance between the lens and the agar surface. System calibration is menu-driven and automatic, and typically takes around 5 min after power-up.
2. Reliable "connectivity" algorithms to rapidly scan the computer image for closed regions that may be individual colonies. These algorithms

Figure 2. General view of our flat-bed picking robot. In the foreground is the light table for illuminating two large culture trays containing the genetic libraries randomly spread on to agar. Behind this fixture are the racks of twenty-four 384-well microtitre plates into which the picked clones are inoculated. The camera and picking 96-pin picking head is moved over the unit by the three-axis gantry robot.

start from a predefined threshold across a dark–light transition. If such a threshold contour can be closed on to itself, then the connectivity algorithm has located a "blob" which is then passed to the next stage of image processing. One major difficulty in this approach is that a threshold value will change across a large colony tray due to slight variation in light intensity from image frame to image frame. In our systems, variation in light intensity across the agar tray is accounted for by an automatic threshold-modification function.

3. Sorting functions to select from all "blobs" located by the connectivity algorithms only those which have visual characteristics of an individual and separated colony. We use various sorting parameters to distinguish individual colonies from touching colonies or even artifactual "colonies" caused by agar bubbles, surface defects or lighting effects. These sorting parameters include the blob size, various measures of "roundness" and average grey-scale. Selection by size removes those blobs typically caused by small bubbles, surface artifacts or gross lighting effects. Various measures of "roundness" are used in combination to ignore those blobs caused by touching colonies, especially when three colonies form a "trefoil" shape which is particularly difficult to deselect using conventional measures of roundness alone. The average grey-scale of a blob can be used to select between recombinant and non-recombinant clones using blue–white genetic selection.

We have combined these three principles within a user-friendly interface (Figure 3), which is used in conjunction with highly engineered hardware to provide accurate, reliable and fully automatic colony picking.

Library Picking

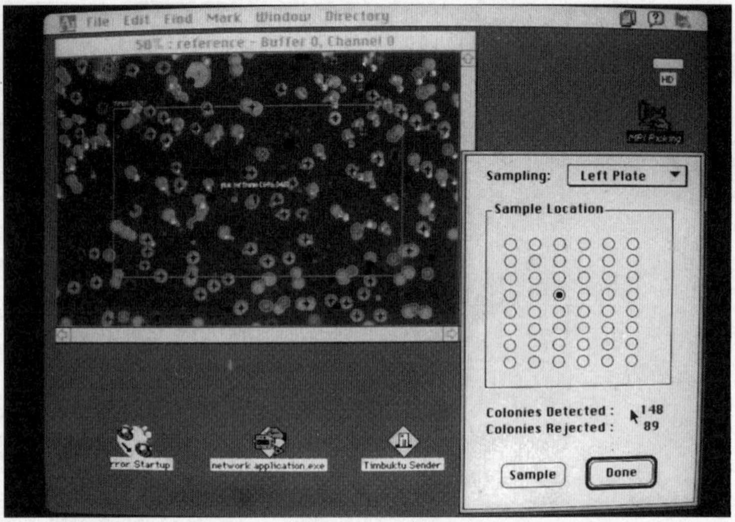

Figure 3. Screen shot of the user interface from the picking robot. The image window represents an area approximately 3 × 5 cm of the agar surface, with blue–white and satellite colonies. The user interface and robot are entirely mouse controlled.

C. Biological Considerations

The construction of genetic libraries is a most critical step, since many months or even years of analysis will be invested in a library. There are several choices to be made at the outset of generating a new genomic or cDNA library which will be familiar to molecular geneticists and include issues such as phage, plasmid, cosmid, BAC, YAC, etc., cloning systems, the amount of rearrangement in the cloning system and the resolution of the library required. However, when automated picking is to be used, additional factors need to be considered for automation to be most efficient. First, the degree of sterility required for a given cloning system. For example, a simple ethanol wash is sufficient to sterilise *E. coli*, but more extreme methods, requiring additional time or robot stations, are required for phage systems. Second, the host strain must be reliable and proven on a large scale, preferably in automated systems.

Finally, it should be realised that no contemporary picking system can pick as reliably as the human hand and eye. If the biologist obtains an understanding of the hardware and software systems that constitute an automated system, it will lead to a greater realisation of which biological steps are most crucial to maximise picking efficiently. For example, to enhance the colony–background contrast for the vision system and to give better growth control over colonies, we typically plate on single YT medium which is paler in colour than the richer 2YT. Media autoclaving should be standardised to reduce colour variation, which will also reduce the colony–background contrast available for the imaging algorithms to utilise. Agar plates of non-uniform or uneven height, if only by 1–2 mm, can drastically affect software conversion of pixel to real-world positions.

We use an automatic pump system to fill a fixed volume of molten agar into colony trays which are poured on a perfectly flat surface. Each agar plate is dried for a standardised time before plating so that agar shrinkage due to evaporation of moisture is uniform.

Plating of the transformation must be conducted with care. We plate approximately 5000–6000 *E. coli* transformants per 22 × 22 cm colony tray. It is important that this high density of clones is plated as evenly as possible for two reasons: (i) to reduce errors in any threshold finding by local concentrations of colonies; (ii) to reduce the number of colonies that touch other colonies. We spread 1 ml of transformation mix using 8–10 small (3–4 mm) glass beads which are shaken vigorously across all directions of the agar surface for 1–2 min. This method produces not only extremely even plating, but also has advantages of speed (since multiple trays can be plated simultaneously), and of not producing colony streaks or scratches in the agar surface which are common when using conventional glass-rod plating tools. Colonies are grown to between 0.5 and 1 mm in diameter and stored at 4°C before picking to ensure the colonies remain dormant during picking.

D. Library Storage and Retrieval

We pick *E. coli* genetic libraries directly into 384-well clear polystyrene microtitre plates (Genetix, UK), each well containing around 60 μl of media. Wells should not be filled more than two-thirds full because on freezing, gas and liquid expansion may push media out of the well leading to cross-contamination.

In large-scale genome programmes, many thousands of microtitre plates will be generated, stored and used in subsequent investigations, which can cause significant storage and retrieval problems. One advantage of Genetix microtitre plates is their relatively low profile which enables many more clones to be stored per unit volume than alternative plates. This can have significant cost implications if plates are to be stored in expensive 80°C freezers. Also, efficient labelling and data-tracking systems must be utilised to ensure that not only can sets of plates be easily located, but that individual plate labelling is consistent and legible. We spray a plate-specific label and barcode directly on to each microtitre plate using an industrial labelling system (Linx, UK). These barcodes are used to track plate movements between freezers and to sample track plates when using our robotic systems.

There are many barcoding possibilities available, and barcoding is too large a subject for a complete discussion in this chapter. However, during design of a sample tracking or laboratory information system (LIMS), it must be decided what information is to be stored in a barcode, and whether the decoded information must be immediately useful or must be related to further information from a database. For example, using a "Code 39" symbology enables both letters and numbers to be encoded but requires more label space to produce a legible code. The "Interleaved 2 of 5" symbology, although considerably more space efficient, can encode only numerical data, has no data-checking characters and can only be printed with specialised printers or software.

◆◆◆◆◆◆ III. PRESENTATION

A. Insert Amplification

For many hybridisation techniques, genetic libraries can be arrayed at high density in the form of *in-situ* clones. These clones are then grown directly on nylon hybridisation membranes, the colonies lysed and processed to fix both the bacterial and cloned DNA on to the filter (Hoheisel *et al.*, 1996). However, for some genome analysis applications (e.g. genotyping, EST mapping, oligo fingerprinting and sequencing), it is necessary to prepare purified insert DNA, and the most efficient method to do this is by thermal cycling.

Many thermocyclers have been assembled, but all large-scale DNA amplification systems are based on water baths as a means of temperature control. For example, water-bath thermocyclers have been built at the Whitehead Institute (USA), at the Berkley Genome Centre and at the Marshfield Medical Research Foundation (USA), where each is used for a different genomic strategy from YAC-based PCR physical mapping to human genotyping.

We use a much simpler version of a large-scale thermocycling robot for DNA amplifications of purified library inserts. We have been able to adapt our reaction conditions in such a way that amplification can be performed directly in 384-well microtitre plates with our robotic thermocycler (Figure 4). Using three heated 225 litre water baths we are able to

Figure 4. Large-scale thermocycling robot designed by, built and used at the MPI Berlin. The basket containing up to one hundred and thirty-five 384-well microtitre plates is cycled between three heated water baths.

cycle up to 135 plates (51 840 reactions) at a time. The basket of plates is moved from bath to bath using a pneumatic X–Z sliding configuration.

Visual Basic software controls the whole system including robot motion, temperature probes and water level sensors. Settings can be changed easily in the user menu allowing different types of DNA amplifications from simple cDNA amplifications to those producing complex products (e.g. Alu-Alu or IRS-amplification) from targets such as individual YAC clones or total genomic DNA.

The polypropylene plates used for thermocycling are heat sealed with a two-sided plastic film and a specially developed heat sealer. The plastic film can easily be removed after the amplification step using the same heat sealer. A variety of plate formats are suitable for heat sealing and water-bath thermocycling and it is important to consider issues such as automation of reaction set-up, reaction volume and plate density before deciding on a particular plate. Suppliers of a range of suitable plates include Genetix (UK), Advanced Biotechnologies (UK) and Robbins Scientific (US).

After the amplification reaction, the DNA product is sufficiently pure and concentrated to be spotted directly on to nylon membranes, in preparation for hybridisation. In our hands, we have found the system to be as reliable as commercial benchtop thermocycling machines.

B. High Density Arrays

Since the introduction of the "high-density array" concept, there is now an immense interest in the use of gridded or arrayed libraries for genome analysis by hybridisation. A number of distinct methodologies exist for creating arrays, including large-scale pin transfer of colonies or DNA (Lehrach *et al.*, 1990), microarraying of PCR products using pins (Schena *et al.*, 1995), piezo-jet delivery of nanolitre droplets (Kietzmann *et al.*, 1997) and photolithographic synthesis of short oligonucleotides directly on silicon surfaces (Lockhart *et al.*, 1996).

Most of these methodologies are subject to different degrees of technical difficulties in construction and proprietorial protection, which may need to be considered even before considering the biological or logistical limitations of each technique. For the sake of brevity, we will concentrate on the approaches we use for large-scale pin transfer of genetic material directly on to hybridisation membranes.

C. Automation of Array Production

Over several years we have been developing gridding robots able to transfer clones stored in microtitre plates on to nylon membranes in high-density arrays (Meier-Ewert *et al.*, 1993; Maier *et al.*, 1994b). Our basic concept is the use of a gantry robot allowing full movement over a spacious robot bed (Figure 5). This arrangement provides greater flexibility and enables us to integrate the picking, gridding and rearraying functions into one robot. Approximately 15 min is needed to exchange robot heads and bed fixtures enabling the robot to be converted to another of its functions.

Figure 5. Large area gridding robot. Up to fifteen 22 × 22 cm filters can be accommodated on to the bed of the robot. The plate stacker containing microtitre plates to be spotted can be seen at the front of the system, the gridding head moves over the bed using the three-axis gantry linear drives.

Robot movement is provided by a three-axis set of servo controlled linear magnetic drives. The entire axis beam acts as the motor's magnet, the drive coils being the carriage, and positional information is provided by an optical encoder running the entire length of the axis. These units provide a high degree of control during sensitive movements, a higher speed (up to 2 m sec^{-1}) coupled with repeatable positional accuracy better than 5 µm. These factors enable the robot to handle precious library plates gently and reliably, and then spot clones at high speed using interpolated movements of three axes.

Various spotting heads can be accommodated by our system. Using spring-loaded devices with 96 or 384 pins of various tip diameters, a variety of biological material can be spotted at different densities. With these machines we routinely spot 15 filters with 57 600 clones in a duplicate pattern in around 3½ hours.

A microtitre plate stacking system at the front of the unit holds 72 plates, enabling the robot to run without any user intervention. A grabber attached to the spotting head of the robot takes the individual plates out of the microtitre plate rack and places them on to a plate holder where the lid is automatically removed and the barcode is read. The barcode reader supplies unique plate identifiers to a database of DNA source libraries and clones, making it easy to locate and retrieve colonies of interest. After the required number of spotting cycles the lid is replaced and the grabber lifts the plate with lid back into the rack system and moves on to the next plate.

For non-radioactive hybridisations in our laboratory using automated image analysis, we routinely spot 57 600 samples on a 222 mm × 222 mm nylon filter (100 times standard 96-well microtitre-plate density). With the system described here we have now achieved experimental spotting densities of up to 147 456 samples per 222 mm × 222 mm membranes, equivalent to 256 times the density of a standard microtitre plate.

◆◆◆◆◆◆ IV. ANALYSIS

A. Hybridisation

Using the gridding systems described above, we transfer genomic libraries on to nylon membranes at a density only possible with accurate robotic positioning systems. Approaching densities of over 10^5 clones per 22 × 22 cm filter means that in one hybridisation experiment, over 10^5 different genetic screens can be conducted. The flexibility of hybridisation as a basic property of nucleic acids allows a broad spectrum of experimental strategies to be applied. Any DNA fragment, from hexamer-oligonucleotides up to YAC clones of more than one megabase, can be used either as probe or as target. Therefore, many different levels of DNA manipulation can be related to one another directly.

Conventionally, radioactive labelling of the probe was used to detect stable duplex formation between probe and target. Currently, there is an increasing shift towards detection methods based on non-isotopic systems. In many cases, however, it still remains faster, simpler and more reliable to use a radiolabelled probe. This is particularly true when only a handful of hybridisations need to be conducted, since many non-isotopic methods require many additional washes or availability of specialised detection equipment. Other experimental limitations may make radioactive detection the only feasible solution, including investigations requiring a highly quantifiable signal and hybridisations using a highly complex or low concentration probe. These particular cases aside, we have developed our hybridisation procedures towards fluorescence detection techniques on nylon filters (Maier *et al.*, 1994a).

B. Non-isotopic Detection

Directly labelled fluorescent probes so far cannot be used as universal hybridisation tools for arrayed DNA filters, although they may be applicable to microarrays constructed on surfaces with low background fluorescence (Fodor *et al.*, 1993; Schena *et al.*, 1996). For broader applications, several enzyme-amplified detection systems are available, for example the enzymatic conversion of BCIP/NBT or chemiluminescent compounds by alkaline phosphatase (AP). However, these systems prove inefficient if many hybridisations per filter have to be processed automatically in a short time and at high resolution. In contrast to the process of chemiluminescence, a fluorescent substrate such as Attophos™ (JBL Scientific, San Luis Obispo) is highly fluorescent after the liberation of the phosphate group, where the number of photons emitted is strongly related to the intensity of excitation light. The signal intensity per labelled probe molecule is much higher than standard fluorescent dyes such as fluorescein or rhodamine, because every active centre of the alkaline phosphatase can process about 10^4–10^5 substrate molecules per minute.

The advantages of Attophos™, which is the benzothiazole derivative of 2′-(2-benzothiazolyl)-6′-hydroxybenzothiazole phosphate (BBTP), in automated DNA multiplex sequencing have been demonstrated (Cherry *et al.*, 1994). We have adapted this system for hybridisation techniques on gridded high-density DNA and *in-situ* colony filters using digoxigenin-labelled probes and their detection via the enzyme-linked fluorescence of BBTP (excitation: 420 nm (max.); emission: >560 nm).

We now analyse fluorescent hybridisations using our purpose built detection system. For BBTP detection, the nylon membranes containing the high-density clone grids are placed in a light proof box and are illuminated with UV (about 365 nm). A high resolution cooled camera is fitted with an interference filter (589 nm, bandwidth about 80 nm) linked to a camera controller and computer. This combination of CCD chip (1317 × 1035 pixels) and fast digitising controller system enables us to rapidly scan a hybridised filter at high resolution. To go from hybridised filter to digitised file takes a matter of seconds, and in terms of sensitivity and resolution our system compares very favourably to commercial fluorescent scanning systems based on photomultiplier detection.

Since the spatial resolution of fluorescent systems can be much higher than phosphor storage screens, we anticipate being able to utilise fully the higher gridding densities capable of being produced by our gridding systems. Therefore, we have implemented other, higher resolution detection systems. These include a time delayed integration (TDI) camera providing a spatial resolution limited only by the linear motion step-size (currently 2 μm), and a laser-scanning system with a 10 μm resolution utilising a sensitive photomultiplier tube for detection.

C. Image Analysis

A crucial necessity for the analysis of high-density arrays is a reliable automated image analysis system able to localise and quantify hybridisation. After testing several commercial systems, we found none had the capability to analyse images of the density or size we produced by our robotic and detection systems. Therefore we developed software to analyse hybridisation patterns of positive signals on high-density hybridisation filters spotted by a robot (Figure 6).

The underlying algorithms of the software package have been developed over 3–4 years and comprise a set of C-programs managed by a shell script implemented in a UNIX environment. The core of these procedures is the grid-fitting routine, a sophisticated statistical method that allows distortions from the ideal aligned grid due to warped filter membranes, bent robot pins, missing spots and high intensity background regions. The grid localisation is performed on a global and on a local scale, in contrast to other methods which normally only utilise local information. Conventional methods search spots first and then fit the grid afterwards –

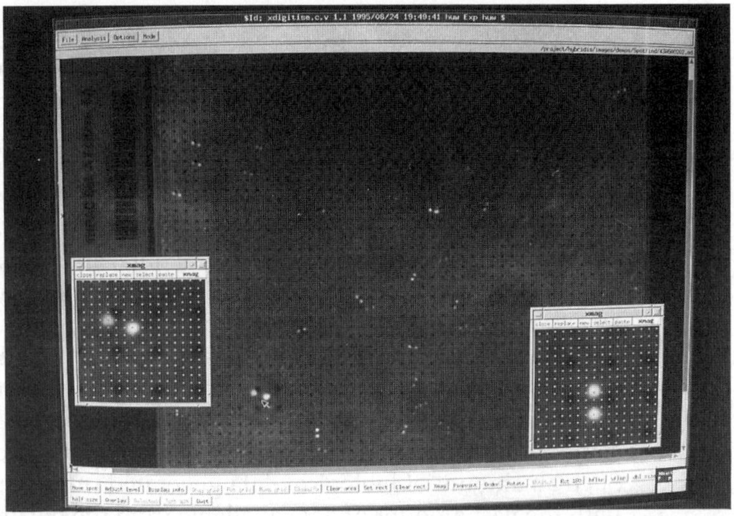

Figure 6. Interface to image analysis software. The screen shows a typical fluorescent hybridisation image, with positives determined from their relative duplicate spotting pattern. The magnified regions show the individual nodes which have been automatically found by the software. The local intensity of every node is then reported to an output file suitable for further analysis.

an approach that often proves non-optimal because it requires that spots are well defined in shape and intensity, and that the image contains few false spots due to background noise.

In this method, spot-node positions are localised by identifying a minimal energy distribution based on a Markov Random Field template, and reflects a trade-off between the regularity of the grid and the trust in the image data. A certain amount of deviation from the ideal grid geometry is allowed, and the grid model is fitted to the image by searching the maximum *a posteriori* estimate through a simulated annealing scheme. This is an iterative sampling scheme from the underlying density, proposed by Geman and Geman (1984) and since used frequently in image restoration problems. The scheme is performed as a number of sweeps over the image, where each grid node is visited.

Once the positions of individual nodes have been determined, the quantification of signal intensity is relatively straightforward. The major problem is of course to determine the factors that decide whether an individual spot is positive or negative with respect to hybridisation and which works on a range of different quality images. Human judgement is so far the only basis by which automated decision making can be compared, but it may not always be reliable. Therefore, we store the pixel values of all nodes of a hybridisation image and perform a global analysis at the end of data analysis. With this approach we can repeatedly analyse the results of large-scale library characterisation projects using various positive–negative thresholds and form an estimate of the confidence limits for our overall results.

D. Bioinformatics

Vast amounts of biological information are now available world-wide. There are well over a hundred DNA sequence, mapping, genetic, structural and biochemical databases available on the World Wide Web (Benton, 1996). The use of computer-based information systems to analyse and integrate these resources with local data is commonly referred to as bioinformatics. The goal of bioinformatics is to find relationships between some partially characterised biological data, and other functionally characterised data held within a global *in-silico* laboratory. For example, if a locally analysed DNA fragment appears to be a gene, a search for identifiable functional motifs may uncover a similarity to other genes of known sequence, structure, biochemistry and ultimately function.

When conducting a genome-scale mapping, gene-hunting or expression study using arrayed libraries, vast amounts of data are generated. Such an amount of biological data generated from such projects, coupled with the vast amount of data available from the *in-silico* laboratory of the Internet, makes highly automated technologies feeding directly into bioinformatic processes an absolute necessity.

◆◆◆◆◆◆ V. THE NEXT STEPS

Analysis and bioinformatic process can electronically identify many interesting candidate clones. However, there is an important distinction between *identifying* clones of interest and then physically retrieving this subset of clones in a form suitable for further sequence, biological or functional analysis. We have set up a robot to retrieve clones of interest from large clone libraries as rearrayed subsets in new microtitre plates.

This function has been included in our existing picking and gridding robots by making hardware changes and writing a rearraying program capable of controlling both the robot moves and the crucially important task of data tracking. Large numbers of bioinformatically defined clones (provided to the robot as a plain text file) are individually and automatically taken from "mother" plates, which are held in the plate stacker, and inoculated into "daughter" plates placed out in the bed of the robot. The robot takes the required mother plate and uses the standard picking head to collect individual clones, changing mother plates when required. Once all 96 pins of the picking head have been used, or the end of the run has been reached, all 96 pins of the picking head are inoculated into a quadrant of a daughter plate. The picking head is then sterilised and the cycle repeated until all user-defined clones have been rearrayed. Full clone-specific sample tracking is provided, using barcode reading, informative user interfaces and extensive data-file output to ensure the location and identity of each rearrayed clone is fully described, checked and recorded.

We have developed high throughput technologies to enable large clone libraries to be picked, presented and analysed – technologies that effectively act to "filter" the huge amounts of genetic data down to smaller numbers of candidate clones, and present these clones to the biologist in rearrayed microtitre plates. The challenge both to the biologist and to the automation specialist is to push the efficiency and relevance of steps beyond this automated "filtering" of genetic information. These next steps will include a full functional analysis of such candidate clones (Fields, 1997) – a challenge likely to need even closer integration of biological, computing and engineering skills.

Acknowledgements

We are very grateful to Sebastian Meier-Ewert, Igor Ivanov, Holger Eickhoff, Markus Kietzmann, Huw Griffith, Martin Horn, Thomas Przewieslik and Sigrid Rumbaum in the Abteilung Lehrach at the Max-Planck-Institut für Molekulare Genetik in Berlin who contributed their ideas to this chapter. We wish to thank Linear Drives Ltd (UK), Acuity Imaging (UK and USA), Genetix (UK) and KayBee Engineering (UK) for their help in developing much of the automation described here and Karsten Hartelius for his collaboration on the image analysis application. The Max-Planck-Gesellschaft and the Bundersministerium für Bildung, Wissenschaft, Forschung und Technologie provided funding for our recent developments.

References

Benton, D. (1996). Bioinformatics – principles and potential of a new multidisciplinary tool. *Trends Biotechnol.* **14**, 261–272.

Cherry, J. L., Young, H., DiSera, L. J., Ferguson, F. M., Kimball, A. W., Dunn, D. M., Gesteland, R. F. and Weiss, R. B. (1994). Enzyme-linked fluorescent detection for automated multiplex DNA-sequencing. *Genomics* **20**, 20.

Fields, S. (1997). The future is function. *Nat. Genet.* **15**, 325–327.

Fodor, S. P., Rava, R. P., Huang, X. C., Pease, A. C., Holmes, C. P. and Adams, C. L. (1993). Multiplexed biochemical assays with biological chips. *Nature* **364**, 555–556.

Geman, S. and Geman, D. (1984). Stochastic relaxation, Gibbs distributions and the Bayesian restoration of images. *IEEE Trans Pattern Analysis and Machine Intelligence* **6**, 721–741.

Glover, D. M. (1984). *Gene Cloning: The Mechanics of DNA Manipulation.* Cambridge University Press, Cambridge.

Gress, T. M., Hoheisel, J. D., Lennon, G. G., Zehetner, G. and Lehrach, H. (1992). Hybridization fingerprinting of high-density cDNA-library arrays with cDNA pools derived from whole tissues. *Mamm. Genome* **3**, 609–619.

Hoheisel, J. D., Maier, E., Mott, R., McCarthy, L., Grigoriev, A. V., Schalkwyk, L. C., Nizetic, D., Francis, F. and Lehrach, H. (1993). High-resolution cosmid- and P1-maps spanning the 14-Mbp genome of the fission yeast *Schizosaccharomyces pombe*. *Cell* **73**, 109–120.

Hoheisel, J. D., Maier, E., Mott, R. and Lehrach, H. (1996). Integrated genome mapping by hybridisation techniques. In *Nonmammalian Genomic Analysis: A Practical Guide* (B. Birren and E. Lai, eds), pp. 319–346. Academic Press, San Diego, CA.

Jones, P., Watson, A., Davies, M. and Stubbings, S. (1992). Integration of image analysis and robotics into a fully automated colony picking and plate handling system. *Nucl. Acids Res.* **20**, 4599–4606.

Kietzmann, M., Kalkum, M., Maier, E., Bancroft, D., Eickhoff, E., Ivanov, I., Przewieslik, T., Horn, M. and Lehrach, H. (1997). Piezo-ink-jet based pipetting-system for high density gridding and nanowell filling. Abstract Automation in mapping and DNA sequencing. EMBL, Germany.

Lehrach, H., Drmanac, R., Hoheisel, J. D., Larin, Z., Lennon, G. G., Monaco, A. P., Nizetic, D., Zehetner, G. and Poustka, A. (1990). Hybridisation fingerprinting in genome mapping and sequencing. In *Genome Analysis: Genetic and Physical Mapping* (K. E. Davies and S. Tilghman, eds), pp. 39–81. Cold Spring Harbor Laboratory Press, Cold Spring Harbor, NY.

Lennon, G. G. and Lehrach, H. (1991). Hybridization analyses of arrayed cDNA libraries. *Trends. Genet.* **7**, 314–317.

Lennon, G. G., Auffray, C., Polymeropoulos, M. and Soares, M. B. (1996). The I.M.A.G.E. Consortium: an integrated molecular analysis of genomes and their expression. *Genomics* **33**, 151–152.

Lockhart, D. J., Dong, H. L., Byrne, M. C., Follettie, M. T., Gallo, M. V. X., Chee, M. S., Mittmann, M., Wang, C. W., Kobayashi, M., Horton, H. and Brown, E. L. (1996). Expression monitoring by hybridisation to high-density oligonucleotide arrays. *Nature Biotechnol.* **14**, 1675–1680.

Maier, E., Crollius, H. R. and Lehrach, H. (1994a). Hybridisation techniques on gridded high-density DNA and in-situ colony filters based on fluorescence detection. *Nucl. Acids Res.* **22**, 3423–3424.

Maier, E., Meier-Ewert, S., Ahmadi, A., Curtis, J. and Lehrach, H. (1994b). Application of robotic technology to automated fingerprint analysis by oligonucleotide hybridisation. *J. Biotechnol.* **35**, 191–203.

Mardis, E. R., Panussis, L., Rifkin, L., Simonyan, A., Stuebe, E., Weinstock, L. A., Wilson, R. K. and Waterson, R. H. (1995). Laboratory automation for high throughput genome sequencing at the Washington University Genome Sequencing Center. Abstract 3rd International Conference on Automation in Mapping and DNA Sequencing. LBNL, California.

Meier-Ewert, S., Maier, E., Ahmadi, A., Curtis, J. and Lehrach, H. (1993). An automated approach to generating expressed sequence catalogues. *Nature* **361**, 375–376.

O'Neil, R., Andre, C., Benson, S., Cassel, J. *et al.* (1997) New chemistries and systems for high throughput automated sequencing. Abstract Automation in mapping and DNA sequencing. EMBL, Germany.

Sambrook, J., Fritsch, E. F. and Maniatis, T. (1989). *Molecular Cloning: a Laboratory Manual*. Cold Spring Harbour Laboratory Press, Cold Spring Harbor, NY.

Schena, M., Shalon, D., Davis, R. W. and Brown, P. O. (1995) Quantitative monitoring of gene expression patterns with a complementary DNA microarray. *Science* **270**, 467–470.

Schena, M., Shalon, D., Heller, R., Chai, A., Brown, P. O. and Davis, R. W. (1996). Parallel human geneome analysis – micro-array expression monitoring of 1000 genes. *Proc. Natl. Acad. Sci. USA* **93**, 10 614–10 619.

Uber, D. C., Jaklevic, J. M., Theil, E. H., Lishanskaya, A. and McNeely, M. R. (1991). Application of robotics and image processing to automated colony picking and arraying. *BioTechniques* **11**, 642–647.

van den Engh, G., Asbury, C., Basiji, D., Dillon, K., Esposito, R., Fey, C., Gelderman, R. and Knowles, S. (1995). Automated clone selection and gel loading for genome mapping and sequencing. Abstract 3rd International Conference on Automation in Mapping and DNA Sequencing. LBNL, California.

Zehetner, G. and Lehrach, H. (1994). The Reference Library System – sharing biological material and experimental data. *Nature* **367**, 489–491.

4 The PREPSEQ Robot: An Integrated Environment for Fully Automated and Unattended Plasmid Preparations and Sequencing Reactions

Gerhard Kauer and Helmut Blöcker

GBF (Gesellschaft für Biotechnologische Forschung), Department of Genome Analysis, Mascheroder Weg 1, Braunschweig, Germany

CONTENTS

I. INTRODUCTION

It is now widely accepted that genome sequence analysis has a major impact on progress in biological sciences and many fields of application, for example in medical diagnostics, plant breeding or drug development. Much of the debate about large sequencing efforts focused on the extent (and hence the costs) of such projects. As in early stages of the human genome project, it is often proposed to restrict future genome projects on cDNA analysis. However, despite the larger costs for total genome sequencing it has been decided at an international level that the human genome with only some 5% of coding sequence will be sequenced completely. This is due to the growing evidence that the non-coding regions of genomes may contain particularly interesting mining treasures. If only cDNA and certain genomic regions were to be sequenced, the sequencing effort and the costs for the genome projects would be immense. We believe that a big step ahead at the technology level is required to reduce the costs and at the same time improve the quality of the data.

METHODS IN MICROBIOLOGY, VOLUME 28
0580-9517 $30.00

There are quite a few approaches to automation of some of the steps of large sequencing programmes. A number of pipetting platforms with adequate software to carry out sequencing reactions in an automated fashion are available from commercial suppliers, some of them with built-in PCR facilities. Other robotic workstations were designed to automatically prepare plasmids from bacterial cultures (Hilbert *et al.*, 1998). Hawkins and colleagues reported on a M13-based automated system they named Sequatron (Hawkins *et al.*, 1997). A few years ago we decided to design a robotic environment which should display the following features:

- Integration into one continuous process of DNA preparation and sequencing reactions.
- Fully automatic and unattended mode of operation to be able to run the system 24 hours a day.
- Non-black box, open design to be able to operate any of the modules separately in cases of service on other modules.
- Common user interface for all the modules for easy operation by technical personnel.
- Easy implementation of new/additional modules by appropriate software design.
- High mechanical stability to minimize teaching frequency.
- High quality of the sequencing results ("best technician grade").

We report here for the first time on details of the design and implementation of such an automatic robotic environment which is now being used in routine sequencing projects. Intermediate reports have been given earlier on a number of national as well as international occasions.

◆◆◆◆◆◆ II. THE CURRENT SYSTEM

A. Overview and Performance

Figure 1(a) gives a photographic overview of the current system, whereas Figure 1(b) is a schematic representation. In Figure 1(a) various valves and the vacuum pump are underneath the top of the table. The waste water line is on the far left. The system as depicted here has been used to prepare samples for three different types of DNA sequencing machines: ABI 377 (Perkin-Elmer, 64 clones), L-4200-L-2 (Licor, 32 clones) and a modified ALFexpress (Pharmacia Biotech, 16 clones). As an example we present a Licor-type protocol in Table 1.

B. Description of the Modules

1. *The desk.* To accommodate most of the modules a desk was constructed (203 cm × 303 cm, height 87 cm). The frame was built from square steel tube (3 cm), the top is 2 cm-thick plastic. The legs were fixed to the floor. Wherever deemed necessary the modules (pipetting platform,

(a)

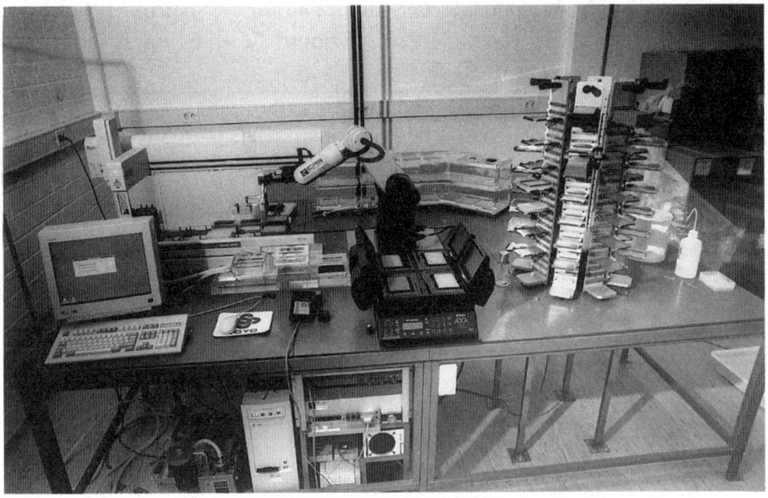

(b)

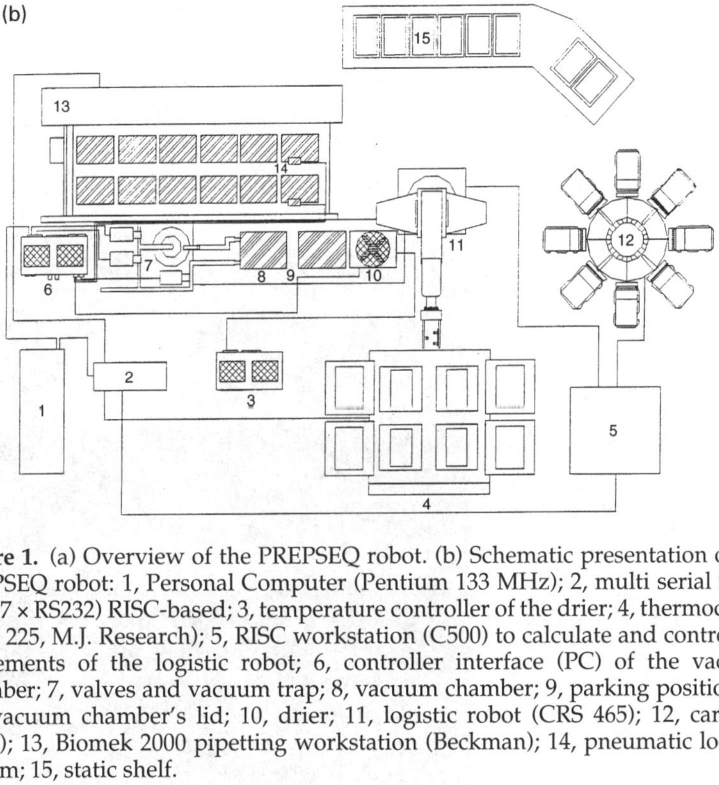

Figure 1. (a) Overview of the PREPSEQ robot. (b) Schematic presentation of the PREPSEQ robot: 1, Personal Computer (Pentium 133 MHz); 2, multi serial interface (7 × RS232) RISC-based; 3, temperature controller of the drier; 4, thermocycler (PTC 225, M.J. Research); 5, RISC workstation (C500) to calculate and control the movements of the logistic robot; 6, controller interface (PC) of the vacuum chamber; 7, valves and vacuum trap; 8, vacuum chamber; 9, parking position for the vacuum chamber's lid; 10, drier; 11, logistic robot (CRS 465); 12, carousel (CRS); 13, Biomek 2000 pipetting workstation (Beckman); 14, pneumatic locking system; 15, static shelf.

carousel, logistic robot) were further immobilised through the top of the desk by legs of their own.

2. *The logistic robot*. This stationary robot is a CRS 465 from Canadian Robotics Systems (CRS). The hand of the robot had to be designed and manufactured from scratch to meet the specific demands of our robotic environment. The global design of the hand is such that the CRS 465

Table 1. Example of a timetable for 96 plasmid preparations and subsequent sequencing reactions (Licor), about 12 × 96 clones can thus be processed every 24 hours

	Time (min)
Resuspension	20
Lysis	5
Neutralisation	5
Transfer to filter plates	9
Pass to Qiaprep plate	1
Pass to waste	1
Washing with Qiagen Washbuff PE	3
Pass to waste	1
Washing with Qiagen Washbuff PE	3
Pass to waste	1
Drying	20
Elution	5
Sequencing reaction	10 (Licor)
Cycle sequencing	45

can easily place labware not only on to the pipetting platform but also into the four-headed PCR machine (Figure 2).

3. *The carousel*. This module is also from CRS. Just as the CRS 465 robot, it

Figure 2. The logistic robot, placing a microtitre plate into the thermocycler.

is operated from the central PC through the same RISC workstation (C500, calculates any movement of CRS 465 and carousel). The commercial trays were replaced by new ones (Figure 3). The main features of these are: (i) (optional) portrait mode as opposed to the old, landscape-only mode trays; and (ii) larger angle between ground plate and the upright parts (120°) to ease correct positioning and removal of the labware.

4. *The shelf*. The current plexiglass shelf is a low cost module to increase the storage capacity, particularly for labware with rather agressive liquids. A number of interchangeable adapters have been created to accommodate various types of labware.
5. *The pipetting platform*. This module was initially a standard Biomek 2000 (Beckman Instruments). Following the experience we gained in the course of our automation project, we introduced some alterations which turned the Biomek into a mechanically very stable module. The main body of the Biomek was tightly connected with four screws to the ground plate, which in turn was directly connected to the floor (see above). This prevented any misplacement of labware on the Biomek by the logistic robot following accidental de-adjustment. Occasionally we observed that the pipetting tools of the Biomek managed to lift a complete box with pipette tips – a disastrous event in any unattended mode

Figure 3. Modified trays on the carousel.

of operation. We therefore fixed the metal plate in positions A6 and B6 to the body of the Biomek and fixed the two labware holders to the metal plate. To keep the pipetting tip boxes tightly in place and to simplify their release we introduced a pneumatic system (Figure 4). On software command it moves the spring in the corner of the labware holder away from the tip boxes, thus circumventing the long-term mechanical problems we encountered with the original Beckman mode of operation.

6. *The vacuum chamber.* Careful design and manufacture of this module were very crucial for unattended plasmid preparation following Qiagen's protocols (Kauer and Blöcker, 1997a). Since any misplacement into the vacuum chamber of the Qiagen trays would severely damage the whole system, four plastic "tongues" were fixed to the inner part of the walls. The tongues will thus guide the trays smoothly into the vacuum chamber even if the logistic robot tries to place them slightly off the ideal position. Unattended and reliable closure of the vacuum chamber is another rather difficult problem. We solved it by carefully balancing the weight of the lid, the selection of the sealing material on the lid and the body parts of the vacuum chamber and finally the applied underpressure. To standardise the DNA preparation, reduced pressure was applied differentially in an upper and a lower compartment of the vacuum chamber. This helped a complete elution from all wells of the upper trays before the elution started from the lower layer. Furthermore, the vacuum chamber was connected to four electronic proportional valves (including an appropriate electronic interface) to provide an automated liquid waste management system and to adjust the underpressure. The vacuum chamber is depicted on the left-hand side of Figure 5 with its lid in the parking position (middle of the figure).

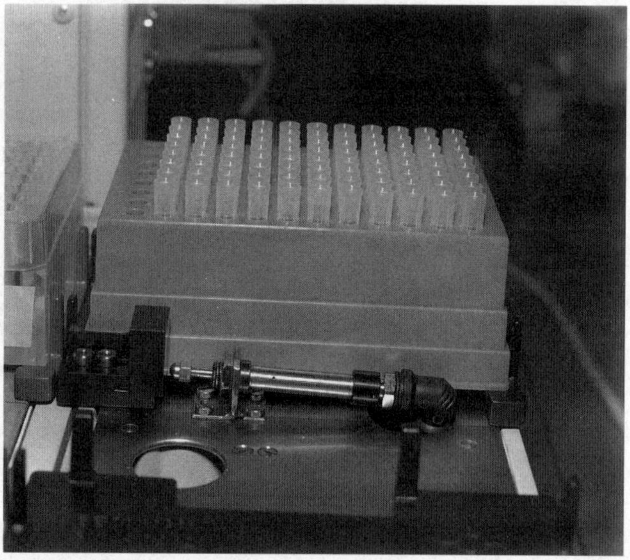

Figure 4. Pneumatic system on the Biomek.

Figure 5. Vacuum chamber, lid and drying module (left to right).

7. *The drier*. The protocol for DNA preparation as distributed by Qiagen involves a washing step of the absorbed DNA with ethanolic buffer. It is known, however, that even traces of ethanol will poison the subsequent enzymatic sequencing reaction. To circumvent this inherent danger of the protocol, residual ethanol from the bottom part of the 96-well Qiagen trays is usually removed by tapping the trays a couple of times on the lab bench which is well covered with tissue paper. The presence or absence of ethanol is detected by smell by the operator. This whole procedure is obviously not perfectly suited for a computer controlled automatic process. We solved the problem (Kauer and Blöcker, 1997b) by the development of a ventilation system which blows warm air (55°C, max.) from underneath against the bottom part of the Qiagen trays, thus quickly evaporating any residual ethanol. A pair of specific tangential fans and the design of the housing help to provide an even stream of warm air. The heating and ventilators are both controlled by the central software through an electronic interface. The upper part of the drying module is depicted on the right-hand side in Figure 5 and a schematic drawing is given in Figure 6.
8. *The software ("virtual robot")*. The software integration of modules from different manufacturers turned out to be rather complicated. We had to deal with different interfaces and also different basic operating systems. We decided to design and implement software which would enable us to deal with communication problems of the present as well as future modules of the robotic environment. The design ("virtual robot") is object-orientated (Rumbaugh *et al.*, 1991) and we used software ICs ("software LEGO") whenever possible for the implementation in C++ (Gamma, 1995). The software is specified after ISO 9001 (ISO92, ISO94). Given that the communication protocol for a new module is known, its software integration will take as little as 1 to 8 hours maximum (Kauer and Blöcker, 1997c).

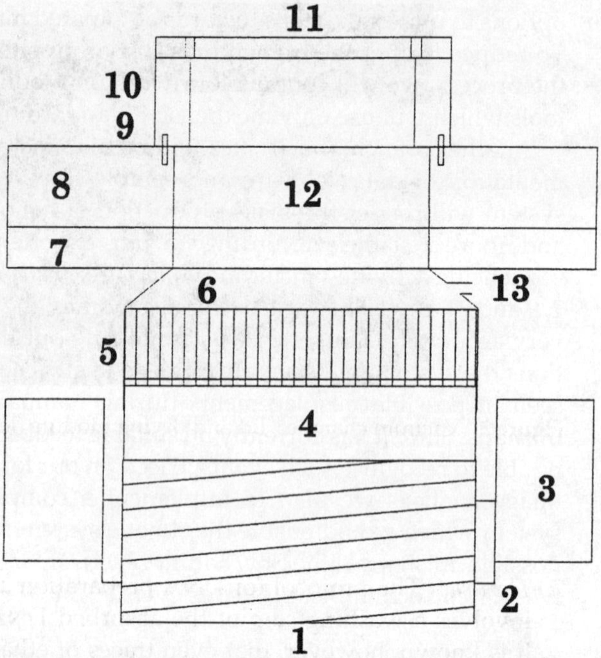

Figure 6. Schematic drawing of the drying module: 1, fan; 2, driving axle; 3, drive motor; 4, opening for the nozzle; 5, heating; 6, nozzle; 7, desktop; 8, ground plate (plexiglass); 9, fixing bolts; 10, spacer; 11, holder for Qiagen filter; 12, air shaft; 13, sensor inlet.

◆◆◆◆◆◆ III. FUTURE DEVELOPMENTS

As pointed out earlier in this chapter we found the system very reliable in its current state of development. We are absolutely sure that without this robotic environment we would not even have the chance to catch up with a number of other international genome sequencing centres. Nevertheless, much remains to be done. Although not an ideal solution, we expect fewer failures by implementing a photometer for online control of the DNA yield. This will enable us to rescue (whenever possible) a certain fraction of too dilute DNA samples by subsequent automatic adjustment of DNA concentrations in the respective sequencing reaction mixes. The general flexibility of the system will certainly benefit from a future integration of a refrigerated cabinet in which any temperature-sensitive biochemicals could be stored and fed into the process in a programmed fashion. We are currently evaluating the implementation of capillary electrophoresis-based multichannel DNA sequencers or equivalent ones, based on slab gel electrophoresis and a novel gel loading system. Such an upgrade of the current system would represent another dramatic leap ahead towards a very competitive high throughput system for genome sequence analysis. According to our estimations, every 24 hours such a system would generate more than one megabase of raw data in a fully automated and unattended fashion, which would offer in turn further options to undertake steps towards automated interpretation of raw data. To reduce further the tremendous cost of pipetting tips and to speed up the process, we will continue our work on modified and novel pipetting tools which will use only inexpensive standard tips.

Another branch of activities will be directed to the control and documentation of data, labware and sample flow. Such an extended LIMS system will render the mass production of genomic data easy to follow and more reliable. According to our experience in various genome sequencing activities, including *Arabidopsis thaliana*, human chromosomes 2, 9 and 21 as well as some bacterial genomes, the current system operates very accurately. However, we cannot rule out that under certain conditions the system will severely damage itself. The major source for this is seen in possible misplacements during manual logistic work or other mishaps. Since it has currently no suitable feedback, the system would not be able to recognise such events. Based on our longstanding experience in image analysis we plan to implement a complex video and sensorial system which would enable the robotic system to deal with a number of possible mishaps either by halting itself or, in the best of all possible worlds, correcting the erroneous event.

Acknowledgements

We thank the German Federal Ministry for Research and Technology (BMFT, now BMBF) and the Projektträger BEO for support for this work through grant 0310703 to H. B. and David N. Lincoln for carefully reading the manuscript.

References

Gamma, E. (1995). *Design Patterns: Professional Computing*. Addison-Wesley.

Hawkins, T. L., McKernan, K. J., Jacotot, L. B., MacKenzie, J. B., Richardson, P. M. and Lander, E. S. (1997). A magnetic attraction to high-throughput genomics. *Science* **276**, 1887–1889.

Hilbert, H., Schäfer, A., Collasius, M. and Düsterhöft, A. (1998). High-throughput robotic system for sequencing of microbial genomes. *Electrophoresis* **19**, 500–503.

ISO92 (1992). Qualitätsmanagement- und Qualitätssicherungsnormen. Leitfaden für die Anwendung von ISO 9001 auf die Entwicklung, Lieferung und Wartung von Software (Identisch mit ISO 9000-3:1991), Beuth Verlag (in German).

ISO94 (1994). Qualitätsmanagementsysteme. Modell zur Qualitätssicherung/QM-Darlegung in Design, Entwicklung, Produktion, Montage und Wartung, Beuth Verlag (in German).

Kauer, G. and Blöcker, H. (1997a). Differential vacuum chamber. Patent pending.

Kauer, G. and Blöcker, H. (1997b). Drying module. Patent pending.

Kauer, G. and Blöcker, H. (1997c). Virtual robot. Patent pending. A simulation program for the virtual robot is available on request.

Rumbaugh, J., Blaha, M., Premerlani, W., Eddy, F. and Lorensen, W. (1991). *Object Oriented Modeling and Design*. Prentice Hall.

5 Building Realistic Automated Production Lines for Genetic Analysis

Alan N. Hale
Oxagen Ltd, Abingdon, UK

◆◆◆

Automated Production Lines

CONTENTS

◆◆◆◆◆◆ I. THE VISION OR MISSION STATEMENT

Laboratory automation projects can fail. Often this is because the scope of the work is not fully appreciated, particularly by the senior management. It is not unusual for an individual to be asked to implement an automated solution to a laboratory process in their "spare time", fitting it in between the so-called real work. Lack of the right budgetary funding also hampers the effort and even a deep suspicion from co-workers as to the motives behind the project causes the developer problems.

The developer is often chosen because he or she is seen to have a flair for computers or instrumentation. This is of course important but the developer should possess a wide range of skills as discussed later in this chapter.

It is essential that the automation project is properly managed and fully specified at the start. A written and agreed vision is useful but it must be widely communicated so that the purpose of the project is understood by both senior management and the laboratory workers.

The style of the vision or mission statement will vary and will be met with a range of responses within different organisations but typically might take the following form:

METHODS IN MICROBIOLOGY, VOLUME 28
0580-9517 $30.00

- To design, develop and implement a fully automated environment for genetic analysis.
- To ensure all processes are integrated in a cohesive and unified system.
- To allow data to be generated and shared so that the process is seamless.
- To provide an environment which allows and stimulates intellectual progress.

A. To Design, Develop and Implement a Fully Automated Environment for Genetic Analysis

This encompasses the whole process. The developer will design a system based on, but not necessarily copying, the manual procedure(s). This will then be developed by buying or building system components and integrating them together. Implementation takes place when the end-users begin to use the system for real work. The phrase "fully automated environment" is a key one. A common error in automation is to assume that it only involves the hardware and software of the system, whereas in reality an automated integrated solution can only work to its full potential when a system is built around it that controls the operator interaction, the sample tracking, the results tracking and so on. The developer should literally create a full environment in which the automation hardware is only one part.

B. To Ensure All Processes Are Integrated in a Cohesive and Unified System

When developing the hardware and its supporting systems, they must fit with existing procedures and processes which are not directly linked with the automated system. In other words the new developments must be compatible with other existing activities. All new development should be done with the broader, global picture in mind. The system must be built for change.

C. To Allow Data to be Generated and Shared so that the Process is Seamless

The operators must not experience any difficulties in extracting the information from the automated system and using it in any way that is appropriate. Once again the way data and results are presented from the automated system must be done with reference to the global picture.

D. To Provide an Environment which Allows and Stimulates Intellectual Progress

Automation in the laboratory differs from, say, automation in the car industry. When manufacturing a car one of the prime objectives is to make reliable cars more cheaply and with shorter manufacturing times. The end product is clear: it is a car. In science there are comparable examples, the end product is a result. However, even if the result is a simple yes or no, it still requires intellectual interpretation. The automation must therefore genuinely free the operator to do other, hopefully rewarding, tasks.

◆◆◆◆◆◆ II. STRATEGY AND OBJECTIVES

The vision or mission statement is important but because of its required lack of specificity the automation project must be defined in more detail before it actually starts. This should not be restrictive in any way to the developer but must define the objectives and the milestones. Without this, the success or failure of the project cannot be assessed. Although this applies to all projects, automation projects often give rise to unrealistic expectations in people's minds because they can be so poorly understood.

The objectives must be achievable and agreed by all those concerned. The developer should be required to report progress on a regular basis and should be responsible to someone in the organisation who can check performance against the agreed objectives. This person should have, or be able to develop, an appreciation of the challenges facing the developer. He or she should be in a position to discuss and agree changes to objectives and milestones as experience is gained with the automation project.

The basic format of a typical automation project is as follows. This varies according to the complexity of the process(es) and the general availability of suitable commercial equipment but serves to highlight the key elements:

- Build information bases on equipment and processes.
- Review commercially available systems.
- Establish internal and external contact networks.
- Perform full evaluations of potential equipment.
- Make recommendations and gain approval(s).
- Implement recommendations.
- Test and refine systems.
- Provide training and operate systems.
- Enhance and refine based on "in-use" experience.

A. Build Information Bases on Equipment and Processes

This process is ongoing almost through the whole lifetime of the project but early and reliable groundwork is essential. The process to be automated, even if the developer is familiar with it, should be discussed with as many people as is reasonable. Listening to people external to the organisation can be very useful in gaining a broader perspective.

B. Review Commercially Available Systems

Wherever possible, the developer should attempt to secure commercially available equipment rather than build a unit in-house. This is typically quicker and makes use of external organisations' greater resources. It also spreads the risk. The supplier has a vested interest in supporting their product and will often provide personnel and practical help. It also relieves the developer from the burden of regular maintenance and repair as this can be delegated to the supplier.

C. Establish Internal and External Contact Networks

The establishment of a network of contacts across a wide range of activities is essential to the developer's successful completion of the automation project. The developer must actively seek expert opinion and be open to ideas and suggestions. An integrated solution to a laboratory process covers many disciplines and developers with expertise in all these varied skills are likely to be rare individuals.

D. Perform Full Evaluations of Potential Equipment

Time must be taken at the early stages of the project to research thoroughly and identify all the commercially available products, especially those in, for example, Beta Testing and close to market. In a larger project the risk should be spread by selecting some well-established and proven technologies but without leading edge technology the lifetime of the integrated solution is likely to be limited. Potential products should be shortlisted and detailed evaluations conducted. All this effort must be fully documented to provide reference material for the future and to allow a hindsight review of the reasons for good and bad purchase decisions. A more detailed purchase strategy is presented later in this chapter.

E. Make Recommendations and Gain Approval(s)

It is important that before committing to a given integrated solution all the recommendations are formally reported (separately) to the senior management and the future potential operators. It is possible that neither group will fully appreciate the subtleties of the design (until they see it working) but without their commitment and co-operation successful completion of the project will prove to be even more difficult. Again full documentation is essential. In larger organisations a formal "sign off" should be considered.

F. Implement Recommendations

In other words build the system. However, time and care should be taken to determine the order of the work. Whenever possible, start with a part of the project that demonstrates a quick and preferably visual success. Usually there will be a lot of interest in the early stages and the right initial impression provides a buffer zone when the inevitable problems start. Remember, if the groundwork has been done properly all problems will have a solution, it might just require a little time though.

G. Test and Refine Systems

Regular testing and re-testing are essential to ensure new or modified ideas have not pushed something out of balance. If possible, the end-users should be involved but choose the assistants carefully. During development it is easy for a system to develop a "bad press" if the operators do not understand that the testing is meant to show where the system fails. The developer must be prepared to refine the system during development as real-time experience is gained. Better to admit that a seemingly good idea at the recommendation stage was wrong than religiously stick to it rather than loose face. At the end of the day the project has to work.

H. Provide Training and Operate Systems

The developer's job should not end when the system is built and tested. Full operator training in use, error correction, routine maintenance and minor repair should be provided. External contacts for support and maintenance agreements should be set up. The developer should provide all the operator training for the first few weeks or months of operation (depending on the complexity of the system). There should be a clear and progressive period of hand-over.

I. Enhance and Refine Based on "In-use" Experience

The developer can learn a great deal about future system enhancements by actively encouraging negative and positive operator comment. As much attention must be paid to what the operator likes as to what is not liked. Rapid response and effectiveness of bug correction during the hand-over period by the developer is critically important. The end-users must have a good feel about the system, want to use it and most of all want it to be successful.

◆◆◆◆◆◆ III. PROCESS FLOW DETAILS

A. Functional Requirement Specification (FRS)

The first step in determining the project specification is to establish clearly the required end result. Often this changes in the light of information gained from future discussions but without it the project specification is impossible to establish reliably. The developer should be aware of embarking on too ambitious an objective. It is better to identify multiple stepwise objectives in order to achieve the ultimate vision, each of which can be produced within a few months, rather than embark on a project which lasts for years before a useful product is seen.

B. System Design Specification (SDS)

1. Process flow

Once the basic objectives of the automation have been established and documented, the process flow details have to be determined in detail. This is typically done by evaluating the manual process. If Standard Operating Procedures (SOPs) exist then this is an excellent starting point. However, there is absolutely no substitute for talking with the people who actually do the work.

The developer should be prepared to take the opportunity to change the way the process is done if the end-users are making these sorts of suggestions. The developer should be acutely aware of the need to hear the tricks of the trade and the shortcuts used by the operators. Often these are not offered automatically, either because the operator instinctively reverts to the way the procedure is supposed to be done or because he or she simply does not recognise that a shortcut is used. Here it is important to talk separately to as many operators as is reasonable.

When all the steps have been identified a block flow diagram should be produced. Often this will be the first time the process has been presented in this manner and it is the only way to ensure that everyone is talking about the same things. An example is shown below. The diagram can be fully professional, using all the correct block shapes to represent decision points, etc., or it can simply be a process list.

Locate sample microtitre plates in freezer storage
Remove from packaging and allow to thaw
Determine sample processing order
Determine process parameters (spot densities, arrangement of duplicates, etc.)
Prepare sterilisation materials
Sterilise tools, clean work area
Supply membrane filters
Apply spots
Remove and store prepared membrane filters
Remove, repack and return sample microtitre plates to freezer storage
Clean work area

2. Procedural requirements

Next the developer should establish the procedural requirements. For example, the precision and/or accuracy of each and every pipetting step must be agreed and understood. The reasons for an operator in the manual process using an accurate dispensing pipette instead of a plastic dropper are likely to be lost in the mists of time and may simply have arisen because of expediency rather than a need for that liquid to be precisely dispensed.

Another example is when the operator may "add acid until the solution becomes frothy" which, although possible to automate, is likely to be challenging and of questionable reliability. In this instance the only solution is to do some basic research before committing to the automation project with a view to establishing a measurement of some sort. In this example, it may be that in fact the same volume is added each time or an excess of acid is sufficient or acid can be added until a given pH is reached.

3. System in-use validations

The system validation points need to be established. These may not all be done in the manual procedure but checks that the automation itself is functioning correctly are very valuable. This can of course be particularly important in an environment subject to inspections from regulatory bodies.

4. Operator interactions

The way the operators work with the system needs to be fully appreciated by the developer. For instance, will the user enter the operational parameters directly or should the system get this information from a database which can be maintained remotely? What type of system alert mechanisms are suitable? How does the operator maintain consumable supplies, how are the samples and the end products taken to and from the system?

Typically the developer will advise the end-user on operator interaction specifications. The developer will normally have a better understanding of how these work (or do not work) and will have a broader understanding of what is possible. However the developer must always remember that they are not the ones using the system routinely and, within reason, should develop the operator interactions as specified by the end users. The customer may not know best but if they do not like the look and feel of the system they will lack enthusiasm for its use.

5. Maintenance requirements

Regular maintenance of an integrated system is essential. Clearly defined daily, weekly, monthly and quarterly, biannual, annual maintenance procedures must be discussed at least in principle. The exact requirements may not be fully appreciated until experience is gained on the system but at least the obvious ones should be covered at this stage.

6. Change control

The documentation behind the Functional Requirement Specification and the System Design Specification is important. It ensures the developer has in writing exactly what is to be achieved. It means the project does not mushroom once people realise the potential and it allows the end-user actually to get a final product rather than one that continually needs just one more adjustment.

However, many automation projects can be regarded as scientific research. As such, new ideas, sudden inspirations and the dawning of the obvious present themselves as the project progresses. It is in the interests of the developer and the end-users to keep these strictly under control but they must be allowed to happen. To allow this, change control procedures must also be defined as part of the System Design Specification.

Change control means all changes are documented and agreed. The reasons for the change must be made obvious and clear benefits must be given and understood.

7. Test documentation

Precisely how the system is to be tested and what results are to be achieved must be agreed. This can be difficult to define precisely and it must be remembered that the automation only has to achieve the same standards as the manual procedure. It is a misconception that automation has to provide an improved standard compared with the manual procedure, although obviously this is desirable.

8. End user training

Ideally at least one end-user will have been involved during the development of the system and will therefore have a good understanding of how things work. However, the developer should specify who will be trained on completion of the project and what form this training will take. For example, some personnel will simply operate the system, some will be able to solve minor problems and others will be capable of more advanced troubleshooting.

◆◆◆◆◆◆ IV. SELECTING THE SYSTEM COMPONENTS

It is likely that the organisation already has a comprehensive purchasing strategy. Typically this will identify an individual with accountability for the purchase. The developer must be in a position to have a major influence on purchase decision making. A corporate preferred supplier can present problems for the developer, restricting the choice of supplier. The suitability of the instrumentation for the agreed automation objective is critical to the success or failure of the project. The developer must be proactive in influencing or preferably making the purchase decisions. The following describes a typical approach to successful procurement of the right equipment at the best costs with appropriate support:

- Assign person with overall accountability.
- Identify and justify the requirement.
- Conduct a purchase review.
- Make the purchase evaluation.
- Determine the purchase decision.
- Make the purchase.
- Installation.
- Post installation.

A. Assign Person with Overall Accountability

A single person with full accountability to authorise the purchase must be established. This person can delegate all the responsibility for the purchase but at all times retains ownership of the purchase decision. The developer will quickly lose credibility if a supplier and a product are identified but then an order fails to be placed and must therefore establish a professional working relationship with the person able to authorise the purchasing.

B. Identify and Justify the Requirement

Before approaching potential suppliers the required function of the instrument must be clarified. This gives the sales representative the opportunity to assess whether his or her product will be able to compete and saves time for everyone concerned. The developer must identify potential suppliers with care and achieve as much choice as possible in the first instance.

C. The Purchase Review

Having created a list of potential suppliers, the developer should first discuss the purchase by telephone and obtain sales literature and list prices (i.e. this would normally be the non-discounted price). Full documentation of contact names and prices with the sales literature must be made for future reference.

A minimum of two suppliers and a practical maximum, typically five or six should then be selected for a more detailed review. This would normally consist of an on-site demonstration by the sales representative. Off-site demonstrations cannot always be avoided but they should generally be discouraged. At this point, formal quotations, possibly with some initial discounting should be requested.

D. The Purchase Evaluation

The list of suppliers from the purchase review stage should be reduced to a minimum of two and a maximum of three or four. If there is clearly one suitable supplier then only this product needs to be evaluated. Common sense and professional judgement should be applied.

An evaluation would normally consist of having the instrument on loan for, typically, 2–10 days. This should be the model and specification which is potentially to be purchased. Sale or return agreements are acceptable but should only be entered into with care.

At least one person should be allowed the time and given the support to fully test and evaluate the instrument. Sufficient space must be allocated and, where feasible, "real" samples should be used.

Table 1 demonstrates a standard template which can be completed for each evaluation. It allows good records to be kept and most importantly easy comparison of different products.

Table 1. Instrument summary

Name	
Supplier	
Application area	Instrument Code:
Cost	
Maintenance cost	
Hardware description	
Footprint	
Arm type	
Arm movement	
Motor type	
Syringe sizes	
Probe type	
Liquid sensing	
Rack types	
Reagent handling	
Stall	
Bar codes	
Software description	
Computer	
Platform	
Development status	
Style	
Consumables requirement	
Customer support	
Training	
Merits and failings	
Ability to integrate	
Personal opinion	

E. The Purchase Decision

For major purchases the intended product might be brought in-house for a second more detailed evaluation.

The following are some guidelines when making the final purchase decision:

- Obtain the best discount.
- Establish the support available:
  - Existence of local servicing facilities.
  - Determine whether service and repair is done on-site or involves a return to base.
  - Determine the number of trained service staff.
  - Establish the opportunity for a loan unit during lengthy repair.
  - Determine the typical response time.
- Determine what training is available and the cost, agree as part of the purchase what training will be required.
- Determine the warranty period and if possible negotiate a longer period.
- Establish, in writing, the nature, suitability and desirability of the service contract.
- Understand and verify the cost of consumables and where they can be sourced.
- Understand and verify any supplies required, e.g. gas, air.
- Check the type(s) of communications connections available and that these are compatible with the other components in the automated system.
- Obtain a list of existing customers and contact a selection to determine how satisfied they are with the product and, critically, with the support.

The above criteria apply to any instrument purchase. An additional requirement, specific to the developer, is a judgement on how well the supplier understands his or her own instrument and how friendly or co-operative the supplier is likely to be when the developer begins to ask for support. The developer's support needs will inevitably be different from and more complex than that required by customers using the instrument in a more "normal" manner.

F. Making the Purchase

This must be co-ordinated with other system components and the unit must not be delivered until the developer is ready for implementation. It is not good practice to make a major purchase and then delay working on the instrument as it can cause concern that the money has not been well spent or that the developer is not in control of the project schedule.

G. Installation

The developer should ensure a delivery date is given and achieved and that adequate space exists and is available on the day of installation. Also, all necessary supplies must be available and tested before delivery, e.g. air supplies.

It is essential that the unit is tested before the supplier leaves the site and full and adequate training is given to all appropriate personnel, most probably this will be the developer.

H. Post Installation

Documented contact names, the price paid, maintenance agreements, etc., should be documented once the purchase is completed. This should also allow checking that the maintenance schedule is being followed by the supplier.

◆◆◆◆◆◆ V. THE AUTOMATED SYSTEM

In this section a basic assumption is made that the automated system has a robot arm to move items from module to module. Other designs include mechanical arms within a module itself and designs which move items from module to module via, for example, track systems. The term module is used to describe any instrument or other units making the automated system.

In this discussion, the robot arm is assumed, regardless of basic design, to have a gripper hand of some description capable of picking up, carrying and placing objects within the working environment.

A. Hardware

1. Robot arm influence on design layouts

Typically a robot arm will move on a linear track of some 1 to 5 metres in length or will rotate about a central point capable of reaching objects within a defined radius of perhaps up to 1 metre. Some examples of layouts using linear track and cylindrical robots are presented in Figure 1(a) and (b). As can be seen, the robot arm has a defined working envelope. All robot accessible items have to be placed within this limited area. Thought needs to be given to the placement of the modules and the positioning of placement areas within the modules themselves. Consideration has to be given to collision potentials between moving parts within the modules and the robot arm. This can significantly affect the layout designs. Also it is not unusual to discover incompatibility of layouts only when items are physically placed on the work area – be ready to modify those wonderfully drawn and thoughtfully planned layouts.

It should also be noted that a typical robot arm has a geometric access envelope, in other words it is rarely a box. Generally the robot arm cannot easily access objects close to the track. The edge of the working envelope is curved, for example the arm can reach furthest at its mid-point. Some arms can access below the track but this requires removing part of the bench top beside the track which brings its own problems. Other types

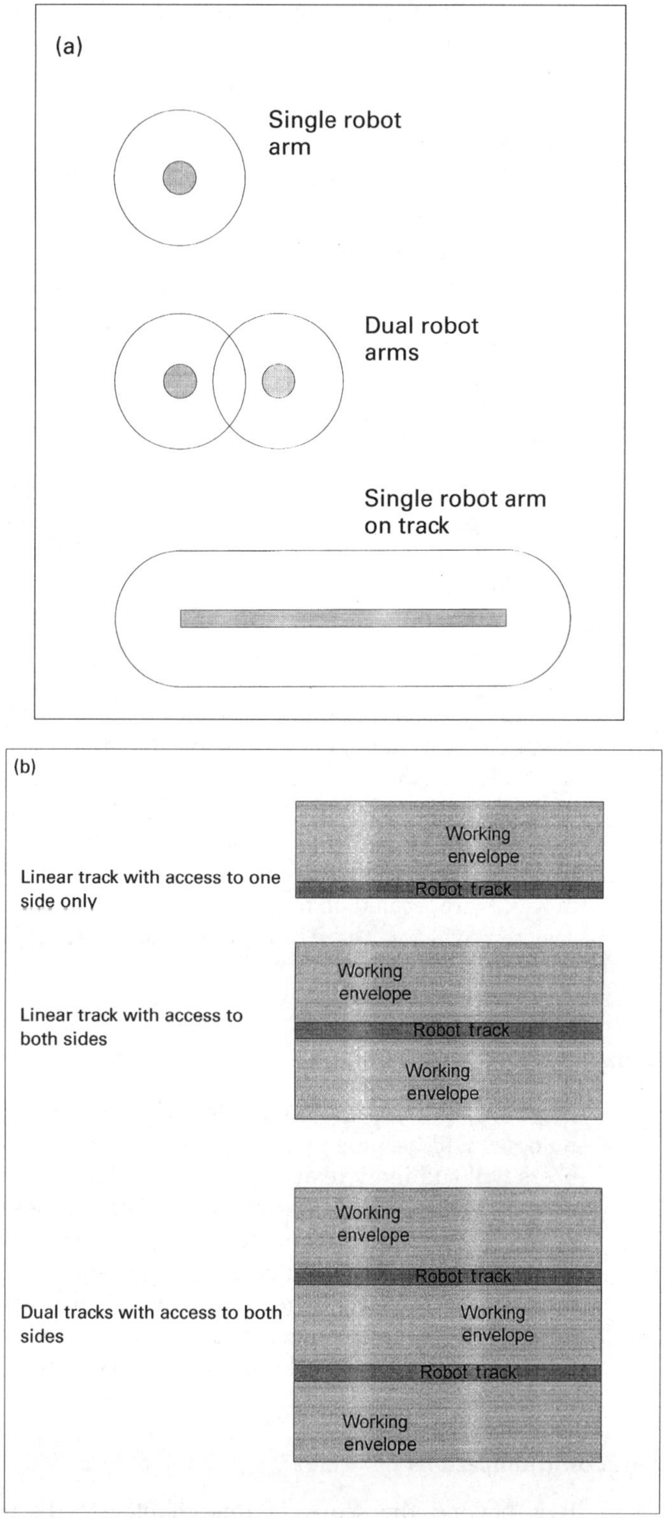

Figure 1. Robot arm configurations showing working envelopes.

allow the track to be suspended above the work surface with the arm hanging down.

2. Three-dimensional designs

Since a robot arm works in three dimensions, the layout can be planned in this way. For example, a large module can be raised and smaller units can be located underneath. Greater robot arm access can be achieved by raising or lowering the track relative to the bench. Typically this would allow objects to be located at the robot arm's mid-point, taking advantage of its point of furthest reach.

3. Operator interaction and maintenance access

Consideration should be given to how the operator will reach positions and components within the system and indeed how the developer is to reach parts within the system during development. Even with the best designed system, the operator will at times have to recover items from areas where only the robot arm has to reach during normal operation. The operator's safety and ease of access must feature highly in the design.

With systems containing modules requiring servicing by external suppliers the service engineer must be able to reach the modules easily and safely. It should also be possible to open the service hatches, etc., on the modules without having to dismantle any of the full system. It is preferable for the service engineer not to have to move any modules. If this is likely to occur on a regular basis thought should be given to how the modules are located on the system and devise ways of allowing quick and easy removal and, more importantly, accurate replacement to avoid major positional re-teaching.

4. Flexibility and components change

Inevitably a successful automated system is going to change. New technologies will become available, modules will be upgraded and the end-users will suddenly realise new potential and ask "but can it be made to do . . .". With the right design at the start, all these changes can be made relatively painlessly. When designing the layout, plan it to change. Leave discrete spaces which can be used to add other components, do not pack modules too closely together otherwise the new model may not fit because it is slightly bigger. Leave access to data points on the modules, they may need to be used if the system expands.

5. System communications

It is beyond the scope of this chapter to discuss electronic communications in any depth; however, the basic choice is between serial and parallel. Generally instrument manufacturers supply one or the other; it is comparatively unusual to see both on a single unit. Therefore the developer often has to use both types on the system. This has advantages providing greater flexibility but can increase the system's complexity.

Controlling instruments via communications ports is not as daunting as a few years ago. With modern programming languages, such as Microsoft Visual Basic™, it can be as simple as putting a control on a screen form and sending the command strings listed in the instrument's instruction manual.

A last resort, which can work well and often adds interest to a system demonstration, is to have the robot arm press the buttons on the instrument's control pad, mimicking the human operator's actions. This lacks sophistication but it does work.

It is also beyond the scope of this chapter to discuss communications protocols such as Object Linking and Embedding (OLE), Dynamic Data Exchange (DDE) and the use of Data Link Libraries (DLLs). Needless to say the developer has to understand these useful features available on today's computers.

B. Software

Typically, the developer will have to work with a range of programming languages on a single automation system, for example:

- High level programming language.
- Machine specific high level language.
- Machine specific control command strings.
- Databases and spreadsheets (e.g. providing data to the system).
- Macros.
- Bar coding.
- Scheduling software packages.

1. Choice of development language(s)

The choice of programming language is partially dictated by the components in the system as well as the developer's own preference. An experienced software engineer may choose C++ and make full use of the advantages it may have over, for example, Visual Basic. In the final analysis however the system simply has to work and work reliably, the developer has to work with sofware he or she can understand and use effectively.

2. Version control and software documentation

In the enthusiasm to create the system, correct version control and software documentation often gets forgotten. This is fine in the short term but as the project complexity grows and time passes, remembering and keeping track of what is what becomes increasingly difficult. Lack of adequate documentation also makes it very difficult, if not impossible, for the work to be transferred and continued by another developer.

System documentation is discussed later in this chapter in more detail. Within the software code some general guidelines are useful, as follows. It should be noted that these comments are intended for the amateur programmer, a professional software engineer is likely to require a more rigid structure.

Each module of software should start with a brief description of its function followed by a chronological list of major changes made with a brief explanation of the change. Minor changes can be documented within the code itself but should be dated. Variable names can and should be descriptive but this should be supported by a brief description when they are defined within the software. Each subroutine, procedure and function should contain a brief explanation of its purpose and its date of creation noted.

Finally, the code should be liberally spread with comment lines indicating what is supposed to happen. It is also very useful to expand these comments when a given routine or procedure proves difficult to work – it is surprising how non-obvious things can be even when only a couple of weeks have passed since it was written.

Providing a version number for each release of software is surprisingly useful. It adds a professionalism and allows for easier discussion of the major changes. This is particularly important when there is more than one automated system of the same design in the organisation. For instance the development system may be on version 4.40 while the main production system may be working with version 4.30 of the software. The version number of the software should be obvious to the user by appearing on the screen.

The following serves as an example of how version numbers may be used:

- *V0.90 series* Pre-release software used exclusively by the developer.
- *V1.00 series* First version of the software to be used by selected personnel other than the developer. Likely to contain many bugs (errors). It is acceptable at this stage to have known bugs which the developer assures the user will be fixed.
- *V2.00 series* The first main release of the software. Essentially free of known bugs it is available for wider general use. This version should work reliably but be used to gain an understanding of what the end-users really want and expect to see the system doing in real use.
- *V3.00 series* The first release of what is essentially the completed software. The experience gained from the earlier releases should mean it is bug free in terms of routine use although some undiscovered ones may yet exist. Version 3.00 series may also be used as a rewrite, tidying up the software, removing the unwanted parts and ensuring that the major items requested by the end-users, as they have gained experience from earlier versions, are provided.
- *V4.00 series and beyond* Each major version change from V3.00 series onwards should represent a major enhancement to the system.

◆◆◆◆◆◆ VI. PERSONNEL

A. Staffing Requirements

As indicated earlier the person selected as suitable for developing an automation system is often one who demonstrates a flair for computers and instrumentation. While this is important, in reality the developer needs to be multi-skilled. This usually means a "jack of all trades" which implies one who is expert in none. As seen from the list below, it is indeed rare to find an individual who is expert in all the required skills.

One solution is to hire a team, each member of which brings a different skills-set. This is a good solution but of course still requires a team leader to provide the management and this individual needs a sufficient breadth of understanding across the whole skills base. Also a team of three to five has a significant financial impact on the cost of the project which may not be acceptable to the organisation.

Another solution is to allow the developer to call on required resources on an *ad hoc* manner. Care needs to be taken to keep costs under control when taking contract workers from outside the organisation but often suitable personnel can be seconded from within. This can cause concerns about career paths for the individuals and can divert effort from their primary tasks, but providing the whole exercise is managed correctly this approach can be successful. The approach taken depends on the size of the automation project and its importance to the organisation.

The individual(s) need to be able to work in a logic-based environment calling on computing, engineering, analytical, teaching, selling, diplomacy and negotiating skills. They need to understand the science of the project, for example, molecular biology, chemistry, etc. Personality traits need to include excellent interpersonal skills, patience, determination and the ability to listen and take advice and criticism. Tenacity is important but so is knowing when to change your mind. The individual(s) must be creative and prepared to take risks but should also be careful and methodical in their approach. Finally, they should possess excellent communications skills to explain highly technical information to others who are not necessarily conversant with automation techniques.

B. Type of End-user

End-users fall into four main types:

1. People who are comfortable with computers.
2. People who are computer literate.
3. People who are not comfortable with computers.
4. People who think they are computer literate.

The order of these categories is deliberate. Type 1 are most preferred by the developer, type 4 are the least preferred. The developer should learn to recognise the types and whenever possible choose end-users of type 1 or possibly type 2 for the initial systems testing (V1.00 series as described above). Type 3 should be avoided until the system is well established with all major bugs corrected. Ideally type 4 should not be allowed on the system but this is perhaps a little extreme. They are the ones who think the developer has an easy task which anyone could do, they often point out the obvious and worst of all can feel confident enough to tinker with the system.

The way in which types 1, 2 and 3 interact with the system should be closely observed. There will be a wealth of information about how to improve it to make it more robust, easier and more obvious (user friendly) to use. Type 4 can stumble across good ideas but their main value is in how to design the software and the system to protect it against accidental misuse.

Developers should always remember that they are building the system for the end-users and not for themselves. You may like screens with bright yellow backgrounds with light green characters but it is unlikely to be appreciated by the majority. The system and in particular the software should be made for the majority of end-users. If time, effort and ability permit, programs can be written which, for example, allow end-users to choose bright yellow and light green but without necessarily imposing it on others.

◆◆◆◆◆◆ VII. SUPPORT INFRASTRUCTURE

When developing an automated system which will have a significant impact on the organisation's productivity it is usually naive to think it will simply replace the existing manual procedures. For such a system to work, careful consideration should be given to the nature of the infrastructure around the system. These considerations include:

- Consumables supplies
- Data input
- Documentation
- Error recovery procedures
- Location
- Maintenance schedules
- Personnel
- Reference materials, sample materials and products
- Results reporting
- Training
- Waste disposal.

A. Consumables Supplies

Consumables supplies include anything used by the automation system which cannot be reused. For example, this might include microtitre plates, disposable plastic liquid handling tips, reagent tubes, ice, system liquids, etc. Procedures to supply and replenish supplies need to be determined.

Critical supplies such as system liquids should ideally be monitored by the system which can then take appropriate action in the event that supply becomes exhausted.

It is useful to determine how the operator wants the replenishment procedures to work. As an example consider the use of disposable plastic tips in a liquid handling robot. The developer writes a program code which tracks the use of the disposable tips and stops the system with a request to the operator to add additional disposable tip racks. This is fine, except when the user has already taken the opportunity to safely replace the empty tip rack holders. The user then has to wait around until the program tells him or her that the tips need to be replaced.

A better solution is for the system to keep track of the disposable tip usage but when it reaches the end of its known supply it should attempt to pick up tips from the beginning again. If the rack is empty then an error is generated and a request for more disposable tips given. However, if the operator has already replaced the disposable tip rack then the system is able to continue without halting. This makes the automation system more efficient: even if the operator is present when the tip replace request is generated it still takes a finite time to complete the operation. Conversely, it does not waste operators' time waiting for the system to tell them what they already know.

As a word of caution at this point, operator safety has to be considered when allowing them access to the system while it is working. This should not normally be a problem but the potential risks must be assessed and the operators made aware and given the appropriate training.

If possible the system should be capable of replenishing its own supplies. Consider the disposable tip situation again. If the system can have access to a stock which it can use to reload the tip racks in the liquid handling robot then clearly this is a better solution. The system still eventually exhausts its supply of disposable tips but it will probably be able to operate without the need for the operator to replenish the disposable tips for a significant length of time.

The developer should consider, and indeed be seen to consider, alternatives to using vast quantities of consumables. The focus of effort should of course be on the more expensive items. Disposable tips, particularly conducting ones for liquid detection, tend to be more expensive than the equivalent ones used for manual operations. The developer might evaluate, for example, techniques for dispensing liquids without using disposable tips where possible. Alternatively, recycling disposable tips might be considered. This would primarily be where the tips were reused to dispense the same liquid rather than attempting to wash and clean the tips to dispense different liquids, unless the risk of cross-contamination was acceptable.

B. Data Input

Getting information to the system from the operator in a reliable and accurate fashion is vitally important. The information that the system needs can range from simply the number of samples to be processed to the dilution factors based on sample concentrations.

The method used should be easy to use and the information entered should be obvious to the observer. As a general rule avoid hidden information, all numbers, factors used in calculations, etc., should be open and easily available to the end-user.

It is preferable if the operator can enter all information at one point, for example on a single system computer. Entering data from an individual instrument's own keypad is to be avoided for two main reasons: first, the operator may forget to make the entry or may make an error; second, the main system probably has no knowledge of the remote data and is therefore operating on blind faith.

The ability to create data remotely and download it to the system computer has distinct advantages but should be used with some degree of caution. It tends to divorce the operator from the system, resulting in the operator losing concentration and forgetting to check something on the system. On the other hand it does allow operators to prepare the information in advance and helps to ensure that they do not accidentally use someone else's parameters.

Barcodes can be very valuable in an automated system. They can be used in a variety of ways but should be more than just an identity number. They should be human readable as well as machine readable. There are many types of barcode and it is beyond the scope of this chapter to discuss them in detail but the two-dimensional codes are well worth considering. This is because they allow more information to be carried per square millimetre than conventional "supermarket type" barcodes and also allow for security coding against accidental damage.

The identity number hidden in the barcode can refer back to a database from where the system gets more information. Or sample processing instructions can be held within the label. The developer can then arrange for the system either to read the barcode and check its identity against a separate sample processing list or the samples can be loaded by the operator in any order and the system can work out this order for itself.

Care must be taken in the design of the operator interface. Elements that need to be checked or changed regularly should be easily available or preferably present on the screen all the time. It is important also that the operator, or the person waiting to use the system next, can at a glance determine progress through the sample list. Ideally he or she should have some indication of how long before the current run will be complete.

It helps if the operator can work with something that looks and feels familiar, for example if data can be entered in the organisation's preferred spreadsheet package. This reduces training and the amount of time an end-user has to spend learning a new package.

Precisely how the operator interacts with the system is critical, it is well worth the developer taking the time and effort to think in detail about how the operator makes these interactions. The developer should never assume the end-user will understand let alone want to learn subtleties of the system.

C. Documentation

Documentation is a very broad subject. Validation documentation is covered later in this chapter; this section is confined to system, Standard Operating Procedures (SOP) and personnel documentation.

1. System documentation

This should be comprehensive and detailed. It is very easy in the flurry of activity to get the system operational to neglect this important area. The following should be recorded for every system component, especially for the modified or custom-made components:

- Supplier name and address. This should include telephone number, fax number, email information and web site.
- Contact names. This includes people's names from technical support, service engineer group and the salespeople. If possible develop a good working relationship with one or more of these people, it helps to have a name to ask for and for that person to know you as the developer.
- Date of purchase. Details of purchase cost, particularly discounts, are useful for future reference.
- Date of installation. This is useful for tracking breakdown and wear and tear frequencies. It should also include a record of any problems which arose and, of course, the solutions.
- Serial numbers. Some companies require this information to be quoted when requesting support but regardless it is easier to make a note at the installation time rather than have to delve into the system once everything has been put together. Also include serial numbers of any additional parts or modules added to the basic instrument.
- Software version numbers and their release dates. Even the best companies supply software with bugs and then correct these bugs in later releases. It is important to know the current version of the software running a given system component. It also helps when building a duplicate system as most, but not all, companies take care to maintain upward compatibility. However, some seemingly trivial change in a supplier's software can quite dramatically affect the developer's software. This is not unreasonable, the supplier cannot be expected to know in detail what is being done with the product. Always ensure full back-ups are made before upgrading any supplier's software.
- List of recommended spare parts with order reference number and approximate price. This saves time when there is a problem. For relatively cheap but critical items it is worth considering a policy of holding stock on site. Some companies will allow this and only charge when the part is actually used.
- Maintenance contract details. Precisely what is covered in a maintenance agreement tends to get forgotten once the negotiations are complete. The developer should not rely on the supplier to keep track of agreements and routine service dates.
- Details of all problems or failures including the resolution to the problem. This is critical and extremely useful to have to hand. It allows experience to be recorded and makes troubleshooting much simpler. With a complex

system over extended time periods it can be difficult to remember precisely how a given problem was resolved. It is also a useful reference point for people other than the developer.

The following list keeps track of how the individual components are connected together and how they interact:

- The units function within the system.
- Communication ports used. This should include technical details, for example baud rate, parity, stop bits, handshaking, etc.
- Supply requirements, for example, three-phase power supply, uninterruptable power supply (UPS), compressed air.
- Consumables requirements including supplier, catalogue codes, cost including bulk purchase discounts.

2. Standard operating procedures

These should be written according to the organisation's internal policies and guidelines but should cover areas such as:

- Routine use
- Non-routine use
- Error correction
- Emergency action
- Safety.

3. Personnel documentation

This documentation area can cover a number of different aspects including:

- Personnel training records. The developer should ensure that only trained and authorised personnel actually use the system. Training by word of mouth should be avoided. Safety issues associated with automatic machinery which may move unexpectedly and at any time need to be communicated and the fact documented.
- Personnel authorisation records. Altering the standard operating parameters may be restricted to certain individuals who are given the authority to make such changes. The changes themselves need to be documented and it should be obvious to all users that a standard parameter has been changed.
- Scheduling of time on the system. In full production use of the automated system, access to the system needs to be tightly controlled to ensure that the resource is fairly distributed and that it is utilised for the maximum amount of time.
- Assignment of responsibilities. For example, routine maintenance may be the responsibility of certain individuals as may be ensuring that system stocks are maintained.

These documents should be constantly updated and modified as soon as anything changes.

D. Error Recovery Procedures

Errors on automation systems fall into three main groups, namely, operator, hardware and software. Errors must be minimised but unfortunately are inevitable in all but the most simple systems. Once an error of some sort has occurred then the most important thing is how it is handled. If an error is handled poorly causing inconvenience to the operator then the automation system will start to get bad press and suddenly the slightest problem will be grown out of all proportion. The problem of course is that until the developer has experienced the types of problems that might occur it can be difficult to predict them and make appropriate preparations. Some potential errors are obvious; many are more obscure.

Operators should be asked to report or document all errors and, in order to encourage this action, the developer must be seen to respond promptly to the reported problems.

1. Operator errors

These are probably the worst kinds of error. The developer has to take full responsibility for all but the most ridiculous of operator errors, as the system should be designed so it is impossible for an operator to make a mistake. This, of course, is almost impossible but every attempt must be made. Operator errors can also be reduced by good training and, as discussed in the section on types of user, by careful selection of who is allowed to work with the system.

Whenever possible the system should test that the operator has set things up correctly. For example:

- Always check that numbers entered are within sensible ranges.
- Do a system check to determine that all required modules are switched on and have been initialised.
- Check levels of system liquids and monitor pressures of gas supplies.
- Perform checks that items such as microtitre plates exist and have been picked up by the system.
- Write to log files so that operator actions or requests are recorded along with system actions.

It should be remembered that the user should feel in control at all times. This is particularly true when, for example, a robot arm is moving up and down the track. This can inhibit the inexperienced user unless he or she feels that the robot is supposed to be moving.

2. Hardware errors

These are generally the most spectacular of error and seem to perform at their best in front of a larger than usual audience. Typically, design hardware errors will manifest during development or when an operator inadvertently leaves something in the wrong place within the system's work space.

Other hardware errors occur when a component fails, for example, a belt stretches allowing a robot arm to continue moving but in slightly the wrong position. These errors can be avoided by regular preventive maintenance and good system design.

Errors like an exhausted disposable tip supply, no liquid in a reagent tube or minor collisions with slightly misplaced microtitre plates fall into the hardware error category and the software has to recognise these and allow the operator to recover without having to restart the program.

3. Software errors

Software errors or bugs take on many forms. Rigorous testing and careful program planning can eliminate most but, particularly with modern style programs where the operator is free to introduce other errors through the software, bugs will remain in almost any program.

All bugs should be trapped by the software even if nothing can be done except terminate the program; the operator should be informed before the whole system crashes.

E. Location

The location of the automated system can be critical to its success. If it is too remote, people are not around to notice if something is wrong or realise in a timely manner that consumables need to be replenished. Alternatively, if it is located too close to people, then noise and heat can become a problem. Whenever possible there should be all-round access to the system from at least three sides. Other factors include close proximity to a drain: it is more convenient for waste to go directly to a drain rather than rely on an operator to empty waste containers.

A typical system can generate quite a lot of heat, as there tends to be a higher than normal density of electrical equipment per square metre with an automated system compared with a typical manual laboratory. Adequate ventilation or air conditioning is therefore important. For the same reasons, demands on electrical supplies can be higher than normally found in the manual laboratory so the load on the electrical circuits needs to be considered.

An emergency stop button is usually required and this has to be located with ease of access and consideration has to be given to ease of routing computer network and alarm cabling.

It is generally better not to locate the system where it will be in direct sunlight and usually the processes themselves work better in ambient temperatures that do not vary more than 1 or 2 degrees in a 24-hour period. It should be borne in mind that if the system is running overnight or through the weekend, then the ambient conditions may change significantly as there are no personnel in the building to complain if it is, for example, getting too cold.

F. Reference Materials, Sample Materials and Products

Once operational, one of the biggest advantages that an automation system has over the equivalent manual operation is consistency. Once programmed, the automation system will tend to do things in exactly the same way time after time. The great advantage the competent human operator has over the automation system is the ability to "see" and "feel" when a process is not quite right. The only way, with today's generally available technologies, for the automated system to "know" if something is wrong is to provide it with reference materials or standards of known properties.

Reference checks should be built into the system's operation. With a high productivity automation system it is amazing how much useless work can be generated before it is realised that something is wrong if adequate checks are not put in place and monitored. The key word here is monitored. The procedures must exist to check routinely the performance of the system and alert when and if things start to drift out of synchronisation.

Getting the sample materials to the system reliably and maintaining sample integrity while on the system is an area that can be overlooked. From an automation point of view it tends to be easier if containers of materials are already opened. Having to unscrew caps or pierce seals can be difficult and unreliable. The downside of this of course is evaporation. It is quite easy to loose samples or see concentration changes over periods of time. By the nature of automation processes, materials tend to be out on the bench for longer periods of time than when being used in the equivalent manual procedures.

The same problems apply to products that are made by the process: they can suffer from evaporation or become contaminated by being open to the elements for too long. All three materials, reference, sample and product, can degrade if not properly maintained.

There are a whole range of solutions to these problems, including, for example, refrigeration, covering with inert materials such as mineral oil, sealing with foil, etc.

G. Results Reporting

Any results generated by the system need to be reported and, most importantly, reviewed in a timely manner. The quicker a poor result is identified the better. Equally, the faster good results are reported the greater the confidence in the system's reliability. This is relatively easy when the system generates and evaluates its own products. In many cases though, the automation process prepares materials for later analysis on remote instruments. It is with these processes that care needs to be taken to ensure post-automation work is done effectively and with a turn-around time that is appropriate. The developer should not be afraid to slow down the automation process to balance correctly the operator's ability to keep up. As stated earlier it is remarkably easy for an automated system to generate vast amounts of product before it is realised that something is wrong. This can be a large waste of materials and resources and leads to bad press for the system, even when, as is usually the case, the system was working with substandard materials.

The way in which results are reported and the presentation is worth thinking about in some detail during the planning and the implementation of the system. Ease of interpretation is of prime importance. Additional system checks, beyond what might have been done manually, should be considered and where appropriate included. It might be a tedious process manually but relatively simple for the automation system to add checks to the process. The developer should take advice but not be discouraged from adding these checks by the end-users who generally will not see the need.

H. Training

A well thought out and designed training program is worth the effort. Training should cover the following topics.

1. Safety

Safe working practises are the most important feature in any laboratory and automation systems are no exception to this rule. The impact and penetration hazards associated with moving robots should be highlighted to operators and non-operators alike. Robots can and do move suddenly and unexpectedly. The developer should remember that he or she will have a better instinct for when a robot is likely to move because of the familiarity with the program code. Therein of course lies the greatest danger – familiarity. This applies to the experienced operator as well as to the developer and there should be regular updates on safety issues for the automated system.

During the design, all options for safety interlocks should be considered and implemented. Emergency procedures should be thought out in advance and form part of the system documentation. Advice should be sought from the suppliers of pieces of equipment and modifications made if appropriate and necessary.

Operators should be clear in their own minds on how to stop the system. By default this is by hitting the emergency stop but there will always almost certainly be other ways of stopping the system. This can cause confusion as hitting the emergency stop probably means a non-recoverable abort of the run in progress and often there is a reluctance to take what can be an extreme measure. The developer should provide clear guidelines on which abort method to use and when it should be operated.

2. System set-up

With any automated system there will be a standard way of initialising the whole system. This includes the power-on order of the various components. Even if it does not matter, it instils the right discipline in the operator to have a set power-up routine. Once the system is powered then the various components may need to be initialised. Wherever possible, this should be done from the central system computer rather than from the remote instrument and again a set order should be suggested. The developer should ensure that the initialisations position components in their safe home positions, which are designed so that they do not restrict operators' access to the instrument and they avoid collisions when the system is started.

Next the operator should perform check routines that evaluate system liquids, the condition of waste containers, the status of consumable supplies. Now is the time to replenish all stocks. After this the operator can begin to supply the system with samples, standards and reagents.

Finally, the developer should stress the importance of that final glance around the work area. It is so easy to leave a lid or empty box in the way, or not position a sample quite right. This final check is critical to a successful automated run and should be encouraged.

3. General operation

Training for general use includes how to instruct the system to process each sample, how to start the system and how to stop or pause the system. The meanings and relevance of all the on-screen options should be explained even if that particular operator is not going to use all the options. The software should be designed to allow the user to experiment, to change parameters but then easily change them back again. This is standard software design but some end-users will be nervous of using the automation and should at least feel comfortable with the software.

4. Routine preventive maintenance procedures

Preventive maintenance takes many forms: some routines will be daily, others weekly and yet others monthly. Probably operators of the user type 2 described in section VI.B above will be selected to perform these procedures. Again full training and the justifications for the effort should be given. All preventive maintenance done should be documented.

5. Advanced operation

Whether advanced operation features exist depends somewhat on the complexity of the system but might include, for example, calibration routines. In the early lifetime of the system this will almost certainly be done by the developer but as soon as practical this should be delegated to end-users who have the right ability and understanding of the automation. The actions and results of the calibration should of course be logged and recorded for future information.

6. Troubleshooting

Most end-users will learn to troubleshoot the system, indeed there will probably be some troubleshooting done that the developer does not know about. End-users are often reluctant to document and therefore admit the more trivial problems they have and solve for themselves but the developer should make every attempt to keep on top of this and be seen to provide fixes for such minor problems.

Ideally, a limited number of key personnel should be shown by the developer some of the more advanced and obscure troubleshooting. This allows some faults to be corrected in the developer's absence.

7. Minor and major repairs

Carrying out of minor repairs is best kept to the developer and perhaps an assistant. It is unwise to let too many people loose on the system with spanners and screwdrivers because this creates a lack of control. Minor repairs could include replacing drive belts and syringe valves.

Major repairs would perhaps be replacing whole robot arms or drive motors. This is best left to the service engineers or to individuals within the organisation who have attended formal and approved service training courses provided by the suppliers.

I. Waste Disposal

Safe removal of waste materials, for example plastic disposable tips, should be incorporated into the organisation's existing waste disposal policies. Automation systems can quickly generate large amounts of waste which should not be allowed to build up in and around the system.

◆◆◆◆◆◆ VIII. OPERATIONAL PARAMETERS

Unfortunately, laboratory automation projects can and do fail. This may be due to many reasons:

1. The objective might be impossible to achieve.
2. The personnel involved may lack the necessary skills to complete the project.
3. The project might be under-funded or under-resourced in some way.
4. The planning for the project might be at fault.
5. The organisation's culture may not want the project to succeed.

Sadly, more often than not, automation projects fail simply because no one knows, or there is no common agreement, as to the end-point of the project. The end-point includes what the system produces, the quality the product needs to reach and in what time span. In fairness, it can be difficult to judge these targets at the beginning of the project when so little might be known about the potential of the automation. The end-user wants the best possible, the developer does not want to commit to targets which might prove to be too difficult to achieve and one of the key elements of a successful automation project is flexibility. The ability to move the goal posts as the project develops and experience is gained, particularly by the end-users, greatly improves the overall success of the project in the purest sense.

A solution to this is to set almost arbitrary targets at the start and review these regularly. However, great care must be taken not to change the targets so often and by so much that the project never reaches a conclusion. When controlled properly, this approach allows the developer to keep the project targets at achievable levels and allows the end-user to identify other potentials for the system. The option to recognise a good idea but note it for a phase two project should be allowed, thereby a target can be set and reached and then a new project identified.

Alternatively, the project can be defined in a highly detailed fashion, with time and effort taken by integrator and by end-user to understand fully all aspects of the process(es). This can then be followed and exactly what was specified can be produced. Providing the project specification process is done well and enough time is spent, then this approach works well and gives a high degree of professional security to the system's integrator. The risk of course comes from trusting that the project accurately specifies what the users require and assumes that the pre-planning identifies each and every problem and eventuality.

There are two key issues that should be understood concerning this section, namely:

- Cost–benefit ratios.
- Measuring automation (metric comparisons with manual approaches).

A. Cost–Benefit Ratios

Broadly speaking a cost–benefit ratio analysis is designed to tell you what you get for your money given a particular approach. Understanding the added value is extremely important, but it can be counter-productive to "bean count" an (ultimately successful) automation project. To put this in context, the development of a new method of analysis for a drug compound may be allowed to take place without critical cost evaluation, providng it stays within a broadly defined budget. This is not to say development chemists are working with an open cheque book but the control and concern is directed at the quality and reliability of the method and the time point at which it becomes available for general use. The reagent costs during development, salary and overhead costs, etc., might be absorbed into the drug development program. These costs are monitored within the organisation and steps taken if they are getting out of hand but essentially they are seen as a necessary part of the drug development process.

Buying a new piece of instrumentation requires more formal justification. Often the need is identified a year or more ahead and if the request is deemed to have been justified then the opportunity to purchase appears in the next budget. The purchase has to be justified by demonstrating the work that is not being done because the resource does not exist, or that the work can be carried out more effectively, etc.

Increasing headcount in the organisation generally requires even more effort and justification and understandably so; new employees cost significant amounts of money and are there to stay. As a result of the justification the decision to employ a contract worker might be taken, the point being of course that the recruitment can be justified and that justification can be understood by the decision makers.

Typically, justifying an automation project is extraordinarily difficult. This is perhaps not the case in all organisations but all too often the developer is swimming against the tide. There are many reasons for this and each organisation will have its own issues but often the developer faces the following problems:

- Large-scale automation tends to be expensive.
- The benefits of the automation are not clearly understood.
- Previous automation programmes may have failed.
- Existing equipment may become redundant as a result of the automation.
- The development time scales are too long.
- How the system will be supported in the future may not be clear.
- Bespoke components may be required.
- The necessary range of expertise may not be generally available within the company.

1. Large-scale automation tends to be expensive

A substantial capital outlay can be required just to get the automation project started. Capital purchases often simply involve installation, a few days of training and setting up and then the equipment is up and running. This differs from automation projects when it can be several weeks before any return on the capital outlay will be seen.

The best response to this is to do the arithmetic and determine at what time point the automation project will break even relative to the current manual procedures. The argument has to be made for the hidden savings, such as freeing operators to do other things, which can be notoriously difficult to estimate. The developer should take care not to bias the calculation with his or her enthusiasm for the project. It should always be remembered that such calculations can be and are retrieved from the filing cabinet at a later date for review. In the ideal situation, when the cost evaluations are made in retrospect, the automation system should be seen to be providing cost benefits in addition to the ones identified during the cost calculations.

2. The benefits of the automation are not clearly understood

It can be difficult to assess precisely what the organisation will gain by automating a manual procedure which might well be working efficiently.

The project justification has to explain why the organisation should take the risk of spending substantial capital just to change the current working practices. Ironically, the justification is not any easier even when the manual procedure is not seen to be working effectively, the question to be addressed then becomes why would it work any better with the automated solution than it does manually.

This relies on the groundwork and image of automation that the developer creates. Backing the words with demonstrations of success helps a great deal and this is where the developer's interpersonal skills are most required. Once again the developer should avoid ambitious claims which can be remembered at a later date and used to demonstrate apparent failure.

In order to achieve this, the developer should have a thorough understanding of the processes involved. He or she should be aware of the misperceptions of the benefits of the automation that will almost certainly exist in other people's minds and should not allow these to become the expectation which fails to materialise.

3. Previous automation programmes may have failed

This problem can be almost insurmountable. Bad historical experiences with automation projects are usually remembered vividly. A new attempt requires detailed justification and rightly so; at the end of the day no automation project should fail if it is properly planned and co-ordinated.

The only solution is careful and diplomatic "selling" of the project. Here more than ever small steps, each demonstrating success, are important. It becomes gradually easier as confidence in the project grows.

4. Existing equipment may become redundant as a result of the automation

This is always a difficult issue to overcome, particularly when the existing instrumentation is comparatively new. The developer can look for ways to incorporate existing items into the automation project or to demonstrate that the existing equipment can be used alongside the automation thereby further improving productivity. If existing equipment is to be used, the impact of removing it from general use while the development is ongoing should be assessed.

Accurate project planning is of particular value in this instance. If possible, the bulk of the automation should be done before the existing equipment is utilised, even for testing. The idea is to "drop" the existing equipment into the system and for the whole system to be operational almost immediately afterwards. If possible, the developer should design the system such that continued use of the existing equipment in the manual fashion is possible, developer and end-user sharing time. When the target times are set during the planning stage, allowance should be made to allow for restricted access by the developer. Bad press of "I couldn't do my work because the equipment was not available ..." is to be avoided.

5. The development time scales are too long

Sometimes, the automation project can take so long to complete and be ready for use that it is quicker to recruit additional staff and equip them to run the project manually. This may well be the case but project extensions or the potential of further new projects requiring the same resources should be taken into account. If this happens then it is possible that it is worth taking the time needed to complete the automation project. Once again the developer should endeavour to give the end-users access to the system to support parts of the manual process as things move to complete automation.

6. How the system will be supported in the future may not be clear

This should be considered very carefully. It is a valid concern and is addressed in the way the system is designed, the supporting paperwork and the training given to end-users. If the developer is working essentially alone then concerns that all the information resides in one person should be addressed; equally, when the development is a team effort, care should be taken that the system knowledge is focused and not spread too widely and that any one person does not hold critical knowledge that is not available to the whole team. This is not any different from any other type of project but, for some reason, can seem more important to an organisation when that project involves automation.

7. Bespoke components may be required

Bespoke components or items made specially for the system cause concern for the following reasons:

1. It is not known whether they really will work until the cash and time has been spent on the development. By the time it is clear that the bespoke component is not going to work, it is often too late in financial and time terms.
2. By their nature bespoke components may only be made once or twice. Getting another one made at some point in the future can prove difficult and getting an existing one repaired may be impossible. It will almost certainly be more expensive to manufacture a bespoke component externally but, with a third party working to current quality standards of documentation, etc., is likely to be able to provide adequate support in the future.
3. When purchasing custom-made components, the purchaser has to pay for all the development costs, which, with a component for a larger market, are spread across many buyers. This tends to make even relatively simple items seems expensive. The developer should have a view on wider commercial exploitation of the bespoke item, in the interests of price and support as well as any potential royalties arising from the idea.

At the end of the day, the developer has to take some calculated risks when recommending bespoke components. He or she should only work with a reputable manufacturing company and should develop good working relationships with that company in order to ensure close co-operation. The additional costs have to be justified by the project and potential wider markets should be discussed and if possible identified with a view to spreading the developments costs.

A reputable company with a close working relationship can be relied upon to keep accurate drawings and documentation about the component. They will be willing and able to carry out repairs and generally eager to manufacture further units.

8. The necessary range of expertise may not be generally available within the company

An automation project of almost any size requires a wide mix of skills. The developer may be expert in one or two of the skills but will usually look to others to supplement areas where strength is lacking. Ideally, these will be found within the organisation; if not then external assistance has to be sought. This can be difficult and the developer also looses some degree of control. The cost and resources of subcontracting the work necessary should be clearly understood before the automation project gains final approval.

B. Measuring Automation (metric comparisons with manual approaches)

With any project it is important to determine targets and goals that are measurable. This is especially important with automation projects because, as has been said before in this chapter, the expectations of the automation project can be poorly judged and understood by others, particularly senior management. This can result in undesirable consequences for the developer and be a disaster for the approval of future potential automation projects.

The following are some of the key points to be addressed. These apply to all automation projects, others will of course apply depending on the nature of the project and the culture of the organisation.

- Is the automation expected to increase the number of samples processed in a given time period?
- What hardware resources are going to be demanded by the automation system compared with the manual process?
- Is the automation expected to operate an extended day?
- How much operator interaction is expected?
- Is the automation project expected to use more or fewer consumables than the manual process?
- Are results expected to be of equal or greater accuracy/precision compared with the manual method?

1. Is the automation expected to increase the number of samples processed?

First, what is actually achieved manually must be determined. Usually this is not as simple as it sounds. It is not atypical for a team leader to err on the ambitious side of target claims primarily for obvious reasons. The developer should take care to allow for holidays, sick leave, training courses, meetings, tea and lunch breaks, etc., when estimating precisely what can be achieved by the manual process. Most importantly, the developer should establish the number of successfully processed samples as opposed to the number attempted as these two values may well be significantly different. It should also be made clear exactly what can be sustained by the operator. An experienced operator may well be able realistically to achieve, say, 200 samples a day, and keep this pace up for 5 days or more but will he or she then collapse in a heap and be off for 10 days to recover. Reality checks should be made at all times and the developer should not forget that extensively used automation systems can suffer from mechanical breakdowns from time to time.

A simple way to establish the sample throughput achieved is to determine the numbers processed by an experienced highly competent person and the number for a trained but relatively inexperienced operator. A per day or per week value can be calculated and then an annual figure based on a 42-week year can be worked out. Allowing for only 42 weeks in a year gives a not unreasonable allowance for the factors mentioned, which include holidays, sick leave, etc. The given organisation may have a standard weeks-per-year figure which can be used instead but the developer should be free to choose to use a value which provides a better fit with automation systems.

Depending on the difference between the two numbers from the highly experienced and the less experienced operator, a realistic target of, say, the mean value can be agreed. There is no reason to accept the "personnel best" figure as a basis for the target, the automation system will be working in the real world.

The above assumes a straight one-to-one comparison of a single operator and one automation system. Consideration should be given to the numbers of operators needed to produce a given target and whether the automation system is to achieve that target. One problem for the developer at this stage is that he or she may not have a clear idea of the required operator interaction with the system and therefore cannot easily estimate the amount of operator time freed when using the automation system. This point is covered later, but must be included in the target calculations. The classic post-development example is when the work produced by a team of ten is used to demonstrate that the manual system is faster, when during the design and development phase only two operators ever worked together on the manual procedures.

2. What hardware resources are going to be demanded by the automation system compared with the manual process?

Unfortunately, the reality is that laboratory automation systems can stand idle. This, ideally, is a consequence of their own success, idle time being used to allow, for example, data interpretation to catch up with the sample production. Other reasons include: insufficient samples to occupy the system full time; a lack of staff to run the system, etc. The developer should be aware that the finger can point at "under-used resource" by others who have no need of the automation but can make good use of some of the individual system components. This can be extraordinarily difficult to handle with diplomacy. On the other hand, automation systems can be tipped out of balance if other operators are allowed to use components in a manual fashion, but equally so, organisations cannot afford to allow capital equipment to stand idle.

The solution is to design manual operation into the automation system. This is good practice anyway, as it allows greater flexibility, for example, allowing the operator to perform quick, one-off experiments. This avoids the inherent slowness caused by waiting for automated systems to work through all the process steps for just one or two samples. Automation systems by definition generally work better with larger sample numbers. Equally so, it allows non-automated use of selected system components. The developer must be aware though of the obvious potential hazard of someone attempting to use the system manually without realising that the automated system is in operation. In the best case the system attempts to use a component which is already in use; in the worst case the manual operator enters the automated system's environment and suffers injury or death. It is indeed true that automation systems can move suddenly and without warning as is often stated on the warning signs.

While the design of the system is under way, the developer has to decide how much capital equipment is required. For example, will the system perform with greater efficiency if two liquid handling robots are used instead of having just one handle all the liquid manipulations? There is a balance to be found between cost and efficiency or even convenience of operation. Another example might be whether to invest in disposable tip recycling capability, which increases the capital cost but may reduce running costs. The dilemma of course is that the savings in disposable tip costs are "hidden" (unless the unit is added retrospectively when the savings can be demonstrated) whereas the capital cost is very obvious to those asked to approve the expenditure.

Once the targets for the system have been established, the developer can then begin to develop a feel for the hardware/equipment resources required. With experience this becomes easier but getting it wrong can be a major problem. If too little capital expenditure is requested then asking for even more at a later date can be a significant issue to resolve. However, overestimating is definitely a major problem and unfortunately tends to be very obvious to almost everyone in the organisation.

3. Is the automation expected to operate an extended day?

Typically, extended operation of automation systems is expected. Unfortunately, designing a system to operate for long periods in a completely unattended fashion is demanding. This works well with relatively simple systems – a good example is the now standard autosampler in high performance liquid chromatography (HPLC). This excellent technology is today well established and quite routine. The operator spends all day preparing samples, loads the autosampler and goes home for the night, returning the next day to find all the samples processed and the results ready for evaluation. In the early days of this technology, however, the operator could return the next day to find the sample that had been loaded as he or she left the laboratory was in fact the last one for that night – the instant the operator left, something went wrong!

In fairness, it was not always the fault of the autosampler, other system components also failed but the end-result was the same – a wasted day's effort by the operator. Problems also occur during the day when the operator is there to take appropriate corrective action. Often, this is a minor adjustment, the action not even recognised as having been done but crucial to the continuing operation of the system. Once unattended, even trivial problems cause the process to stop. The developer can learn from these events and modify software and hardware to minimise the chance of a given error happening again. As experience grows from working on other automation projects then gradually the robustness of each new system improves.

The developer should be careful when making claims for unattended operation. Certainly, in the early days of operation, someone should be routinely checking the system as often as is practical. A phased introduction to extending the system's working day should be carefully planned and thoughtfully implemented – perhaps starting by allowing the system to run on into the evening and arranging for someone to check on things at a time when everything should be complete. This avoids the "bad press" created by people arriving the next morning to find the system has stopped.

If the organisation works a shift system or a staggered working day then the developer needs to ensure that appropriate key personnel are trained so that someone familiar with the system is present at all times. This is also important when specifying the throughput of the system compared with the manual process.

4. How much operator interaction is expected?

Typically, an automated system is thought of as one where the operator presses a button and returns some eight hours later to find all the work completed. This is usually a misconception in laboratory automation. While not impossible to achieve, it is much more likely that an operator will be required to replace consumables such as disposable tips, top up system liquids and load more samples, standards or reagent materials during the working day.

The developer should remember that the objective is to achieve an agreed throughput which should significantly reduce the time spent by an operator. It should not necessarily reduce the operator interaction to zero. As far as possible this should be specified as part of the planning process so that the exact involvement of the operator is clearly understood from the start. If, during the design and development process, the amount of operator interaction can be reduced then so much the better. It is not advisable to be seen to underestimate and have to increase operator involvement. Often, the automation justification will have been achieved on the premiss of reducing operator involvement and failing to achieve this is usually very obvious.

5. Is the automation project expected to use more or fewer consumables than the manual process?

A successful automated system achieving greater throughput for less operator effort, by definition, will use more consumables and reagents in a given time period than with the manual operation.

Careful design can actually reduce the volume of consumables but typically many automation systems either use more than the manual operator or even require more expensive materials. For instance, a liquid handling robot with liquid level detection may require special conducting tips. The operator simply looks for the liquid level, the robot has to detect it in some way. Disposable tips designed to conduct when touching the surface of the liquid will almost certainly be more expensive than their non-conducting counterparts.

An operator can easily compensate for irregularly sized containers; a robot may find this more difficult and require components to have the same dimensions within tight limits of variation. Again, these components may cost more to purchase.

Unless programmed with sophisticated error action procedures, an automation system can carry on processing samples even when something is wrong. The equivalent problem would alert the operator and he or she can then take corrective action before materials are wasted.

The developer should attempt to specify clearly the consumables costs during the system specification period. Any potential additional running costs should be justified. This may simply be a case of identifying the added value of the system which makes the added costs more acceptable. The potential savings to be made based on operator salaries should be used with care. This can make the operators wary of the motives of the automation project and can be difficult to demonstrate in the real world.

6. Are results expected to be of equal or greater accuracy/precision compared with the manual method?

As a general rule, the automation system should only be expected to achieve the same levels of accuracy and precision as the equivalent manual process. It should avoid inter-operator variability and might generally be expected to be more consistent overall. Once again the developer should ensure that no false expectations are raised during the system specification period. It is another common misconception that automation systems are more accurate or precise than the equivalent manual procedure.

6 Examples of Automated Genetic Analysis Developments

Alan N. Hale
Oxagen Ltd, Abingdon, UK

CONTENTS

I. INTRODUCTION

This chapter discusses two major automation projects: namely, genotyping and gridding. These are common techniques in high throughput production line laboratories providing support to leading edge genetic research. In automation terms, they represent very different styles. The genotyping application essentially mimics the manual process and indeed can be done entirely manually. The gridding project, at the defined grid densities, cannot be done manually (at least not at all easily) and here the automated solution has no alternatives.

There are many automation solutions to these techniques, those detailed here are recognised as only one approach and serve to provide examples of successfully implemented projects.

II. AUTOMATED PRODUCTION LINE GENOTYPING

A. Process Flow Details

1. Functional requirement specification (FRS)

(i) Business objective

The Genotyping Group wishes to automate genotyping such that some 20 000 genotypes are processed per week on average. The genotyping group's main objective is to produce validated genotype data ready for conversion to allele numbers for linkage analysis. This is achieved by plating out DNA, adding a primer reaction mixture, covering the reaction mixture with mineral oil, performing PCR thermal cycling and finally pooling (combining) the PCR products. These pooled reaction mixtures are then separated by electrophoresis.

Sample materials are held in 96-well microtitre plate formats, although 384 formats are likely to be required in the near future. The key issue is successful completion of PCR to produce sufficient amplified product for analysis. The main problems are that the reagents used are very expensive and PCR reactions can be sensitive to subtle changes in conditions resulting in complete failure to amplify product for no immediately obvious reason.

It should be noted that the target capacity of 20 000 genotypes per week is a limit imposed by the capacity to analyse the data. As techniques improve in handling the data then increased genotype capacity will be required.

(ii) Project scope

The scope of the project is:
Review of the following approaches:

- Semi-manual
- Semi-automated
- Automated.

This assumes that using an entirely manual-only approach is to be rejected and, procedurally, the following will be needed:

- Specification of a liquid handling robot
- Specification and selection of microtitre plate storage and manipulation
- Specification of primer mix reagent storage and manipulation
- Sample tracking and verification
- Provision of easy to use software capable of flexible application of the system
- Multiple options for individual or a combination of process steps.

(iii) Current manual system

A genotyping project for a high throughput laboratory can range from 200 individuals to over 10 000 individuals. A project size might be expected to be around 800 individuals: with about 94 individuals plus two controls per microtitre plate this represents eight master DNA plates. A current marker set for PCR reaction might consist of some 400 primers resulting in a project of the order of 400 multiplied by 800, i.e. 320 000 genotypes.

The 400 or so markers are divided into panels of several markers, the PCR products of which can be combined (pooled) and run simultaneously on an electrophoresis gel.

The process involves creating duplicate daughter plates of DNA from each of the master DNA plates. A given marker from a panel set is added to each whole daughter plate (typically there are 14 markers per panel and therefore each panel requires 14 daughter plates). Either mineral oil can be added to each well of the microtitre plate or the technique of using a heated lid to cover the thermal cycling block can be used to prevent loss of reaction mixture by evaporation during PCR. In PCR, the reaction mixture is heated and cooled in a precisely controlled manner up to temperatures above 90°C. Often this is done with instruments but water baths can be used, allowing multiple plates to undergo PCR simultaneously.

After completion of PCR, which can take up to 2 hours, the products are pooled or combined by taking samples from under the mineral oil and mixing them in new 96-well microtitre plates. Three pools are taken and then a final microtitre plate containing a combination of all three pools is made. This is called the "pool of pool plate" or the "loading plate". It is this final microtitre plate that is loaded on to electrophoresis gels for separation and subsequent data analysis. (See section II.A 2(i) for a diagrammatic representation of the process.) It should be noted that there are many variations in approach and this description simply represents one alternative. A well designed automation solution is capable of meeting the demands of different approaches.

At a capacity of 20 000 genotypes per week this results in a 16-week project without allowing for repeats due to PCR failures.

(iv) Proposed system requirements

The system will require the operator to supply microtitre plates containing stock solutions of DNA already at PCR working concentrations. Microtitre plate lids will not be used. Tubes containing reaction mixtures for PCR will also have to be provided, most likely in cooled storage racks. It is expected that the operators will standardise on a specific manufacturer of microtitre plate and because of block designs of the thermal cyclers the dimensions of these plates will not generally be subject to change once the component parts have been selected.

The number of PCR microtitre plates to be processed in a run will vary from one to more than 80 and the operator will want to be able to perform the full genotyping process or any sub-part of that process, for example just the pooling steps.

Additionally, the exact time taken for a PCR reaction will vary depending on the conditions of the reaction and will also take different times depending on the age of the thermal cycling block, the ambient conditions, etc. This will have to be taken into account during the programming of the system.

It is expected that the operator will supply already prepared primer mixtures. The system will not be expected to prepare these *in situ*.

The orientation of the microtitre plates must be known but simply has to remain consistent throughout the process. A default orientation will be established to avoid potential confusion and barcodes used for microtitre plate identity will be fixed to a defined end to allow the system to check orientation. If the barcode is not present, the plate can be turned around to see if it can be read from the other end before an error is reported.

Full error handling will be developed. As the system will be designed to be self-scheduling, taking into account variations in PCR reaction times, different panel set sizes, etc., the error handling routines will have to be relatively sophisticated. It is likely that because differing operators will want different responses to a given error, e.g. no microtitre plate present, the system response will be held in configuration files.

For maximum flexibility, the system will be self-scheduling, keeping track of the PCR blocks available at any given time as well as the current capacity of all other system components. Additionally, allowances will automatically be made for the different numbers of microtitre plates in each panel and the different pooling strategies from panel to panel.

(v) Audit trail

A full audit log will be maintained for each run. The number of microtitre plates processed, the process requested for each microtitre plate, the PCR conditions used, volumes used, errors found, operator action taken, start time and end time will be recorded.

(vi) Security

The level of security will depend on the final choice of system components. Appropriate measures to prevent accidental change to system parameters will be taken. No protection to the audit files is envisaged. It will be the responsibility of the end-users to ensure adequate procedures are followed by correct training and the use of appropriate standard operating procedures (SOPs).

(vii) Back-up

All software, parameter files and audit data will be backed up on the local area network using established in-house procedures.

(viii) Training

Selected personnel from the end-users will be trained and will take responsibility for producing SOPs consistent with laboratory policy. The system's integrator will provide the initial training, guidance on safety and assist in the production of the operational SOPs.

With a system of this complexity, specific end-users will become expert in troubleshooting. Often, very minor problems will arise which if handled correctly will allow the process to continue without delay or having to restart the system. How to deal with such problems in a relatively complex system is not difficult but will require experience and a full understanding of the system and its component parts.

2. System design specification (SDS)

(i) Process flow

Locate stock DNA plate(s)
Aliquot to stock DNA microtitre plate(s)
Determine number of daughter plates per stock DNA plate for each marker panel
Prepare required number of PCR reaction mixes
Determine number of pool plates required
Establish sample processing order and PCR program names
Aliquot DNA to daughter plates
Add PCR reaction mixture
Perform PCR on thermal cyclers
Pool PCR products
Prepare pool of pool loading plate
Remove and store PCR products for potential repeats
Clean work area
Run electrophoresis
Collect and process data
Report data

(ii) Procedural requirements

There are many approaches to the automation of PCR. Examples include:

- *Semi-manual.* The operator handles all the microtitre plates and prepares all the reaction mixtures. Typically, an 8 or 12 channel hand-held pipette would be used in conjunction with an instrument capable of aspirating and dispensing into 96 wells simultaneously.
- *Semi-automated.* The operator handles all the microtitre plates and prepares all the reaction mixtures but the liquid dispensing and aspirating are done with a liquid handling robot, which is capable of working with many microtitre plates in a given session. Additionally, the operator will be following a series of instructions provided by the computer controlling the liquid handling robot including information on which PCR program to use, etc.
- *Automated.* The operator prepares all the reaction mixtures but all microtitre plate manipulations and the liquid dispensing and aspirating are done entirely automatically using robot arms and liquid handling robots. PCR programs are started automatically and the whole process from plating out the DNA to the final pooled product ready for electrophoresis is undertaken without any operator intervention, once it has been started.

Each approach offers advantages and disadvantages but if reproducibility, low staff numbers, operator morale and high throughput at low cost are required then the fully automated solution is likely to be the best option. Subsequent discussion will assume the fully automated option is accepted.

The procedural requirements include: reproducibility, flexibility to run small and large numbers of samples with unattended operation and the ability to perform PCR and its associated steps with the minimum number of constraints imposed by the design of the automation system. Microtitre plate order and sample integrity have to be maintained and the use of the correct reagent guaranteed.

Cross-contamination must be eliminated, therefore disposable tips rather than washable probes are likely to be used. The cost of the disposable tips can be reduced by careful and appropriate recycling for pipetting the same solutions if required. At the end of the process disposal of all used tips will take place.

(iii) System in-use validations

- *Database.* Due to the number of genotypes to be processed in a working week (likely to be in excess of 20 000), a database will be required which maintains details of all DNA samples, the reagent and PCR programs to be used, the pooling strategies and other operator-configurable information.
- *Confirmation of sample identities.* Materials identity, for example microtitre plates of DNA, will be checked using barcodes which will be cross-checked with the information in the database.
- *Correctness of operator-supplied materials.* During routine operation of the system, it will have to check for the presence, absence or insufficient quantity of liquid materials, e.g. DNA samples or reagents.

  The system will be expected to confirm the presence or absence of microtitre plates at each stage of the process (not only to check the operator has supplied them but also to confirm that the robots have correctly retrieved and placed the plates).

  The system will also need to be capable of checking for the presence or absence of disposable tips for liquid handling. Even with disposable tip recycling (i.e. reusing for the same liquid within a given system run, high throughput will require significant numbers of tips to be used and supplies will have to be maintained.

(iv) Operator interactions

The operator interface should of course be as simple and easy to use as possible. The genotyping process involves a number of individual steps, which present the additional difficulty that they can and do change from run to run and even from cycle to cycle within a run. There is considerable opportunity for operator error in setting up the run details. Therefore, changing the run conditions must be as simple and foolproof as possible and, most importantly, it should be obvious to the operator precisely what he or she has requested the system to do.

The current status of the system at any given moment should be obvious and easily understood. Once the system is running, it will be self-scheduling and the exact order of processing will be dependent on the run set-up. In other words, often one run will have a different "look" or "feel" to the next one. The operator should be able to check easily what is happening and gain reassurance that everything is proceeding correctly.

The default settings should be the ones the operator is most likely to use and starting, pausing and aborting the process should be clear and easy to follow and enact.

(v) Maintenance requirements

- Daily
  - Tidy and, as necessary, clean work areas.
  - Initialise robot components and run positional accuracy checks.
- Weekly
  - Clean work areas.
  - Run calibration procedures on robot components.
- Monthly
  - Run calibration procedures on robot components (with experience this may replace the more frequent weekly check).
  - Run validation checks on system performance.
- Quarterly or biannually
  - Carry out routine preventive maintenance checks on all major components.

(vi) Change control

- Changes identified during development
  - The end-user group will nominate at least one individual to work with the developer, assisting in providing materials and running tests. This individual will also be responsible for providing proactive input into the working design of the system. Such input will be documented and implemented either at the developer's discretion or after referral to the head of the group. Alternatively, it will form part of a future enhancements document which will detail potential changes to be made after the initial development is completed.
- Changes identified after validation and hand-over
  - All changes to the system once it is in routine use will be documented on change control forms. These will be traceable to the source code which will contain logs of the changes making it easy to determine where and how program code has altered. Any necessary changes to SOPs will also be documented on the change control forms. Details of system tests and results arising from evaluating the change will also be recorded.

    The software will be fully version controlled and this will be updated with each change to the source code. This will be reflected in the appropriate SOPs.

(vii) Test documentation

Once the system components have been identified, the testing requirements will be decided. The testing will include:

- Correct sample flow, e.g. the correct microtitre plate goes to the correct locations at the right time and in the right order.
- Correct selection and implementation of PCR programs and pooling regimes.
- Reliable self-scheduling.

(viii) End-user training

This will depend on precisely which system components are selected but it is expected that all end-users will be trained to operate the system and solve minor problems. All will be able to perform daily, weekly and monthly maintenance checks. One or two will be selected to deal with more advanced troubleshooting and be expected to contact suppliers either to troubleshoot by phone or to arrange for engineers to visit.

B. Selecting the System Components

1. Assign person with overall accountability

For this automation project, accountability will be taken by the systems developer with support from several experienced end-users including the group leader.

2. Identify and justify the requirement

From the groundwork done as described earlier, the system will comprise the following key areas:

- Storage
  - Microtitre plates.
- Robot arm and track
  - To move and manipulate items such as microtitre plates within the system.
- Liquid handling robot
  - To perform all liquid handling steps except mineral oil additions; this includes DNA, master reagent mixes.
- Thermal cycling units
  - To perform PCR.
- Mineral oil dispenser
  - To add mineral oil to 96- or 384-well microtitre plates.
- Barcode printers and readers with labels
  - To produce barcode labels to uniquely identify microtitre plates, allowing these barcodes to be read during system operation. The labels will be resistant to storage at 4°C and thermal cycling temperatures of up to 95°C.
- Sample tracking database(s)
  - These are likely to be developed in-house using commercially available database software.

- Control software
  - This software will be developed in-house, co-ordinating and linking to the commercial packages controlling the individual components. The development language will most likely be Microsoft Visual Basic™.

3. Conduct a purchase view

Each of the broad category areas defined above can be evaluated and reviewed independently. The developer needs to oversee the process to ensure compatibility between the items which he or she will eventually have to integrate into a whole. In many instances, some of the decisions will have already been made within the organisation. For example, barcoding may already be in use and the new system will simply use the organisation's standard. Choice of robot track is likely to be based on experience or influenced by existing systems.

4. Make the purchase evaluation

Full evaluations of all equipment were conducted. It is beyond the scope of this chapter to provide details and indeed product developments continue to move at such a fast pace that similar purchases being made 18 months later are likely to result in different decisions.

5. Determine the purchase decision

As with good purchasing practice, thorough discussions and evaluations concerning training, warranty periods, support, etc., were established.

6. Making the purchase(s) and installation(s)

Purchases were made in a co-ordinated fashion. The strategy adopted with this project was to buy the liquid handling robot and use it in a manual fashion initially, working with the supplier to develop the unit and its software to meet the needs of the group. Once this was completed, the robot arm, thermal cyclers and the storage systems were purchased and implemented.

Barcoding was developed separately to this project and once the automated genotyping operation was implemented, barcodes were added. Concurrent to this was the database development to maintain details of samples, etc.

C. The Automated System

1. Hardware

(i) Liquid handling robot

During the development discussions, a high percentage of the time was spent deciding the requirements of the liquid handling robot. The functionality and performance of this machine was likely to be pivotal to the success of the automation project. There are many instruments available in this highly competitive area which serves to provide plenty of choice but also makes the selection decision somewhat more demanding for the developer.

The instrument has to be capable of the following:

- Disposable tip or fixed probe options and able to utilise a combination of both.
- Eight probes to allow simultaneous single access to a row in a 96-well microtitre plate.
- Variable spacing of the probe allowing access directly into various shapes and sizes of evenly spaced containers.
- Pipetting range of 1–1000 μl.
- Programmable in Visual Basic™.
- Liquid detection capabilities.

As discussed elsewhere, there are many factors which influence the choice of instrument, including product support, cost of consumables, cost of spares and the opinions of existing users.

In order to maximise system flexibility, the disposable tip holders, microtitre plate trays and the other components used within the liquid handling robot were, as is discussed below, supplied as custom-made units.

(ii) Robot arm

A robot arm on a three metre long track was selected to move objects, mainly microtitre plates, within the system. This robot provides high flexibility and ease of programming and easily allows integration with a range of other instruments.

As with the liquid handling robot, there are several competitors available. The developer should work with companies and instruments with which he or she feels comfortable.

(iii) Mineral oil dispenser

The original intention was that the liquid handling robot would dispense the mineral oil used to cover the reaction mixtures in the 96-well microtitre plates. However, during development it became clear that, while perfectly feasible, mineral oil posed a "messy" problem. The solution was to use a commercially available media dispensing instrument, to dispense the mineral oil. This worked successfully and a speciality was the robot arm using its fingers to press the start and stop buttons on the keypad. Later systems lost this feature by using a well filler which allows control via a serial interface.

(iv) Thermal cyclers

The thermal cyclers allow full computerised control via serial or parallel communications. They can be used with mechanical lids but the decision was made to develop a system which utilised mineral oil to prevent evaporation of reaction mixes.

(v) Accessories

The following components were custom made:

1. *Cooling rack*. This was designed to hold the tubes containing the reagent master mixes and keep them at about 8°C.
2. *Tip recycling unit*. Disposable tips represent a fairly significant consumables cost and a high throughput system such as the one described tends to use them in quite high quantities. Tip recycling is programmed to occur when the system determines that it will be dispensing the same liquid again later in the process. In these circumstances, the tips are returned to their rack otherwise they are discarded to waste.

   This has the obvious advantage of reducing costs but also reduces the amount of space required within the liquid handling robot to house sufficient numbers of disposable tips. It has the hidden benefit of reducing the operator need to monitor and restock the system with tip racks during an operational run.
3. *Microtitre plate hotels and robot fingers*. These are generally available components but custom design allows the available space to be maximised for the requirements of the system being developed.

2. Software

Both the liquid handling robot and the robot arm are controlled independently by their own software packages. Therefore Microsoft Visual Basic™ was used to develop a front end application which acts as a user interface and controls the other two packages in the background. The end-user can remain essentially unaware of the presence of the other software.

The sample information is provided in Microsoft Access™ database format or in Microsoft Excel™ format. User parameters and system information is written into a .dat file. This is a simple text file which includes information such as system identity, simulation mode selection (for training and/or software debugging), maximum numbers of microtitre plates allowed, time delays and so on. This can be a .INI file or equivalent but a text file, stored in an application data directory, is just as convenient and is perhaps a little more operator friendly.

Strict version control means Version 0.9x were development copies and the first release was Version 1.0.

The robot arm was taught positions and movements using its own specialist language. The Visual Basic program makes Dynamic Data Exchange (DDE) calls and waits for responses from the robot arm as necessary.

In order to reduce operator effort, the system was designed to be self-scheduling. It keeps track of the space available within the liquid handling robot and the number of thermal cycling blocks currently free. Any one or any combination of the sub-processes, e.g. DNA plating out, master mix addition, pooling, can be carried out at the operator's request and the system determines the processing order. The operator simply specifies the parameters for each PCR plate, e.g. DNA required, reaction mixture to be used, PCR program, pool set, etc., and the system schedules the run.

3. Error handling

From a developer's point of view it is relatively easy to program the system to take a given set of actions following an error, for example a missing microtitre plate. The difficulty is with the end-user who has to decide exactly what action is appropriate. With the missing microtitre plate error the typical options are as follows:

- The system tries again (generally for a set number of times).
- The system stops, displays an informative error message and awaits an operator decision.
- The system ignores and carries on as though the microtitre plate were present.
- The system ignores but moves on to the next microtitre plate.

Typically, error handling is improved following experience of the system in operation. Many errors in the early days of use tend to disappear as the end-users become more experienced and the written operating procedures develop in response to operator errors.

D. Support Infrastructure

1. Consumables supplies

The system uses disposable tips, mineral oil and system liquid (water) with the liquid handling robot. Liquid levels have to be maintained and the system is designed to stop and request more disposable tips when and if supplies become exhausted.

2. Data input

The operator sets up a run by entering the following information for each PCR plate to be processed into a database:

- *DNA plate number.* Up to four master DNA plates can be processed on the system (this default value can be easily changed).
- *Master mix marker tube number.* Up to 96 spaces are available for reagents, these can be selected in any order.
- *Add mineral oil.* The operator has the option to select or deselect the addition of mineral oil. The default is mineral oil addition.
- *PCR program name.* The operator enters each PCR program name.
- *Pooled plate number.* Each plate will belong to a given pooling set, the operator specifies the pool number required.

3. Documentation

Printouts of the system error logs and the sample identity databases are at the discretion of the operator.

4. Maintenance schedules

Maintenance is divided in two main types: operator checks and preventive maintenance. The latter is covered by agreements with the main suppliers. The operator checks form part of the system program. The end-user, following SOPs, runs initialisations and simple positional checks daily. Calibration checks are performed weekly although with experience this may be extended to fortnightly or monthly.

5. Personnel

Initially, two key personnel were trained on the system by the developer and with the assistance of the developer, produced the SOPs and helped to finalise the operational design. After a period of about one month, the process of training all the staff within the group began. After this, all subsequent staff were trained by existing team members and the developer was generally not involved.

6. Waste disposal

Some waste materials are produced by the system, mainly used disposable tips and liquid waste from flushing and in some cases excess reagents. Normal laboratory disposal procedures can be used.

◆◆◆◆◆◆ III. AN EXAMPLE AUTOMATION PROJECT – GRIDDING

A. Process Flow Details

1. Functional requirement specification (FRS)

(i) Business objective

The physical mapping and gene identification group require to automate gridding processes which need to be performed three or more times per week. Gridding involves accurately placing spots of genomic material in precise locations on membrane filters. Ideally densities in excess of 36 000 spots are required on a 22 cm square area. Sample materials are stored in 96- or 384-well microtitre plates and are applied to the membranes using pins – the diameter of the pin is used to control the quantity of genomic material applied. A key issue is contamination, the microtitre plates must not be exposed to contamination. This is because the contents are very expensive and may be in short supply and, of course, contamination will ruin the experiments.

(ii) Project scope

The scope of the project is:

- Specification and selection of a gridding unit.
- Specification and selection of microtitre plate storage and manipulation.
- Sample tracking and verification.
- Maintenance of quality of sample materials, i.e. no contamination.
- Provision of easy to use software capable of providing flexible application of the system.

(iii) Current manual system

The current manual method uses microtitre size filters; genomic material is applied by an operator using a gridding comb with either 96 or 384 pins. The operator dips this into the equivalent sized microtitre plate and applies the material to the membrane. It is impossible for the operator to create membranes with really high densities of spots and therefore the automated system is not only to save operator time but also to significantly increase throughput by achieving higher spot densities.

(iv) Proposed system requirements

The system will allow the operator to load microtitre plates into storage hotels. The plates will have either 96 or 384 wells and will be covered by lids. A range of suppliers are used for the microtitre plates and lids; therefore the system has to be able to adjust to variations in height of plate. The width and length dimensions remain essentially the same from manufacturer to manufacturer.

Typically 48 or 96 microtitre plates will be used per operational run. This may vary. The operator will need the flexibility to start at any plate number, not just the first plate. The system must keep track of this and should prevent the operator from starting with a plate number inappropriate to the process. This will be specified in more detail when the gridding unit has been identified.

It is anticipated that the filter membranes will be loaded manually. This may change when the gridding unit has been specified.

Knowing the order and the orientation of each microtitre plate is critical: any given spot on the processed membrane must be able to be related back to the individual microtitre plate well with complete reliability and confidence. Therefore barcode labels will be applied to each plate, to be read by the system before they are used. The system will orientate the plate based on the end on which the barcode is found. It will be the operator's responsibility to ensure the barcode is on the correct edge.

Full error handling procedures will be developed. The system will check for presence or absence of a lid, a microtitre plate or a barcode. It will be capable of checking the identity of the plate against the expected identity. System responses to errors will be defined in more detail with the end-user, initially it will stop and report the error giving the operator the option to continue, repeat the failed step or abort the process.

(v) Audit trail

A full audit log will be maintained for each run. Start plate, end plate, individual plate identity, errors found, operator action taken, start time and end time will be recorded.

(vi) Security

The level of security will depend on the final choice of system components. Appropriate measures to prevent accidental change to system parameters will be taken. No protection to the audit files is envisaged. It will be the responsibility of the end-users to ensure adequate procedures are followed by correct training and the use of appropriate SOPs.

(vii) Back-up

All software, parameter files and audit data will be backed up on the local area network using established in-house procedures.

(viii) Training

Selected personnel from the end-users will be trained and will take responsibility for producing SOPs consistent with laboratory policy. The system's integrator will provide the initial training, guidance on safety and assist in the production of the operational SOPs.

2. System design specification (SDS)

(i) Process flow

Locate sample microtitre plates in freezer storage
Remove from packaging and allow to thaw
Determine sample processing order
Determine process parameters (spot densities, arrangement of duplicates, etc.)
Prepare sterilisation materials
Sterilise tools, clean work area
Supply membrane filters
Apply spots
Remove and store prepared membrane filters
Remove, repack and return sample microtitre plates to freezer storage
Clean work area

(ii) Procedural requirements

The order of the microtitre plates is critical to the process. It is extremely difficult to trace the true identity of a given spot on a membrane if a microtitre plate has been positioned incorrectly in the processing order. It is therefore essential that the system checks for correct microtitre plate order and identity and only allows the operator to select "logical" sample parameters.

Maintaining contaminant-free materials is also essential. The gridding robot should be capable of reliable sterilisation techniques and will probably include a hepa filter unit. The use of lids on the microtitre plates is essential.

(iii) System in-use validations

During development discussions, the following points were identified as important. Additional system in-use validations will be included as the development of the system and the identity of the system components becomes clearer.

- Positional accuracy of gridding spots
  - The system's ability to position spots accurately will be determined during system validation; in-use confirmation will be restricted to relying on positional feedback.
- Accuracy of microtitre plate locations
  - In-use validation will require operator system checks to determine that materials have been supplied in the correct locations and that any robot

units are correctly calibrated. During operation, presence/absence of microtitre plates will be checked at every manipulation.

- Confirmation of sample identities
  - In-use validation and logging of sample identities will be done using bar-codes. It is expected that the found values will be cross-checked with a database.
- Confirmation of sterile conditions
  - The effectiveness of the sterile procedures will be determined during system validation and this is outside the scope of in-use validation. These checks will be restricted to confirming levels of sterilising solutions, etc.

(iv) Operator interactions

The operator interface should of course be as simple and easy to use as possible. The current status of the system at any given moment should be obvious and easily understood. Full error checking on valid operator inputs must be made, for example the start microtitre plate should be before the end microtitre plate.

The default settings should be the ones the operator is most likely to use and starting, pausing and aborting the process should be clear and easy to follow and enact.

(v) Maintenance requirements

- Daily
  - Clean, not necessarily sterilise, work areas and moving parts.
  - Initialise robot components and run positional accuracy checks.
  - Check for correct operation of sterilising components within system.
- Weekly
  - Clean, probably sterilising, work areas.
  - Run calibration procedures on robot components.
- Monthly
  - Run calibration procedures on robot components (with experience this may replace the more frequent weekly check).
  - Run validation checks on system performance.
- Quarterly or biannually
  - Carry out routine preventive maintenance checks on all major components.

(vi) Change control

- Changes identified during development
  - The end-user group will nominate at least one individual to work with the developer, assisting in providing materials and running tests. This individual will also be responsible for providing proactive input into the working design of the system. Such input will be documented and implemented either at the developer's discretion or after referral to the head of the group. Alternatively, it will form part of a future enhancements document which will detail potential changes to be made after the initial development is completed.

- Changes identified after validation and hand-over
  - All changes to the system once it is in routine use will be documented on change control forms. These will be traceable to the source code which will contain logs of the changes making it easy to determine where and how program code has altered. Any necessary changes to SOPs will also be documented on the change control forms. Details of system tests and results arising from evaluating the change will also be recorded.

    The software will be fully version controlled and this will be updated with each change to the source code. This will be reflected in the appropriate SOPs.

(vii) Test documentation

Once the system components have been identified, the testing requirements will be decided. The testing will include:

- Correct sample flow, e.g. the correct microtitre plate goes to the correct locations at the right time and in the right order.
- Accurate gridding, i.e. the spots appear in neat arrays in the correct locations, equidistant from each other as required.
- Accurate placement, i.e. the right spot is placed at the expected *x*, *y* co-ordinate.
- Adequate sterilisation.

(viii) End-user training

This will depend on precisely which system components are selected but it is expected that all end-users will be trained to operate the system and solve minor problems. All will be able to perform daily, weekly and monthly maintenance checks. One or two will be selected to deal with more advanced troubleshooting and be expected to contact suppliers either to troubleshoot by phone or to arrange for engineers to visit.

B. Selecting the System Components

1. Assign person with overall accountability

For this automation project accountability will be shared between the group leader and the system's developer.

2. Identify and justify the requirement

With the knowledge gained from the research detailed previously, dialogue with potential suppliers can begin. The system is likely to comprise the following main areas:

- Storage
  - Microtitre plate and membrane filters, plus temporary locations to hold microtitre plate lids.

- Robot arm and track
  - To move and manipulate items such as microtitre plates within the system.
- Gridding robot
  - To perform the gridding, taking genomic material from microtitre plates and accurately placing it on to the membrane filters.
- Barcode printers and readers with labels
  - To produce barcode labels to uniquely identify microtitre plates, allowing these barcodes to be read during system operation. The labels will be resistant to storage at –80°C.
- Sample tracking database(s)
  - These are likely to be developed in-house using commercially available database software.
- Control software
  - This software will be developed in-house, co-ordinating and linking to the commercial packages controlling the individual components. The development language will most likely be Microsoft Visual Basic™.

3. Conduct a purchase review

Each of the broad category areas defined above can be evaluated and reviewed independently. The developer needs to oversee the process to ensure compatibility between the items which he or she will eventually have to integrate into a whole. In many instances, some of the decisions will have already been made within the organisation.

Choice of robot track is likely to be based on experience or influenced by existing systems. This is also the case with the storage issues. This project may differ from others because of the thawing requirement of the samples and other existing systems may not have to cope with microtitre plate lids for example.

The gridding robot is the key purchase and this might have to be a compromise between the requirements of the science and such a unit's ability to fit into an integrated system.

4. Make the purchase evaluation

Full evaluations of all equipment were conducted. It is beyond the scope of this chapter to provide details and indeed product developments continue to move at such a fast pace that similar purchases being made 18 months later are likely to result in different decisions.

5. Determine the purchase decision

As with good purchasing practice, thorough discussions and evaluations concerning training, warranty periods, support, etc., were established.

6. Making the purchase(s) and installation(s)

Purchases were made in a co-ordinated fashion. The strategy adopted with this project was to buy the gridding robot and use it in a manual fashion initially, working with the supplier to develop the unit and its software to meet the needs of the group. Once this was completed, the robot arm and the storage systems were purchased and implemented. Barcoding was developed separately to this project. Concurrent to this was the database development to maintain details of samples, etc.

C. The Automated System

1. Hardware

(i) Gridding robot

The gridding robot selected for this project was supplied with six positions for 22 cm square membrane filters, two microtitre input plates and with a hepa filter unit. An access port was added to the side of the unit to allow access by the robot arm.

(ii) Robot arm

A robot arm on a track, able to access both sides, was selected to move microtitre plates within the system.

(iii) Accessories

Microtitre plate and lid holders, microtitre plate hotels and robot fingers were supplied as custom designed and made components.

2. Software

Both the gridding robot and the robot arm are controlled independently by their own software packages. Therefore Microsoft Visual Basic™ was used to develop a front-end application which acts as a user interface and controls the other two packages in the background. The end-user can remain essentially unaware of the presence of the other software.

The sample information is provided in Microsoft Access™ database format or in Microsoft Excel™ format. User parameters and system information is written in simple text file which includes information such as system identity, simulation mode selection (for training and/or software debugging), maximum numbers of microtitre plates allowed, time delays and so on.

Strict version control means Version 0.9x were development copies and the first release was Version 1.0.

The robot arm was taught positions and movements using its own specialist language. The Visual Basic program makes Dynamic Data Exchange (DDE) calls and waits for responses from the robot arm as necessary.

Initially, the gridding robot was controlled by allowing the robot arm to physically press the enter key on the computer keyboard to start the next gridding process. This lacks robustness, as there is not direct feedback that the gridder has completed its work. A timed interval with a margin of safety is used so that the robot arm waited before retrieving the microtitre plates. Future developments supersede this temporary, but effective, solution by allowing direct electronic communication between the Visual Basic software and the Flexys software.

The scheduling with this system was relatively easy: microtitre plates are transferred from storage hotel to the gridder and returned. Microtitre plate lids are removed and replaced. To optimise the process, automatic scheduling predicts when the gridder will finish with the current microtitre plates. From a knowledge of the time taken to remove the microtitre plate lids, the robot arm prepares the next microtitre plates for gridding by removing the lids just before retrieval of the current microtitre plates.

3. Barcoding

The full microtitre plate information is held in a database and this is retrieved based on a simple identity number in human readable form on a label on the microtitre plate. This is supported by a two-dimensional barcode, also applied to the plate which provides useful information when scanned. The operator can read this information (about 64 alphanumeric characters) using a barcode reader without having to refer to the database.

The robot arm moves the microtitre plate to a static barcode reader and confirms the identity number with the one requested in the database. An error is generated if the codes do not match and the system stops until an operator responds. If a barcode is not found the robot arm turns the plate around and tries to read the barcode again; if no barcode is present the system again stops and awaits operator attention by displaying an error message.

4. Error handling

From a developer's point of view it is relatively easy to program the system to take a given set of actions following an error, for example a missing microtitre plate. The difficulty is with the end-user who has to decide exactly what action is appropriate. With the missing microtitre plate error the typical options are as follows:

- The system tries again (generally for a set number of times).
- The system stops, displays an informative error message and awaits an operator decision.
- The system ignores and carries on as though the microtitre plate were present.
- The system ignores but moves on to the next microtitre plate.

Typically, error handling is improved following experience of the system in operation. Many errors in the early days of use tend to disappear as the end-users become more experienced and the written operating procedures develop in response to operator errors.

D. Support Infrastructure

1. Consumables supplies

The only consumable used by this system is the 70% ethanol used in the sterilising bath on the gridding robot. The microtitre plates and lids used to supply the samples are returned to storage after use and the end products, i.e. the membrane filters, are used for experimental purposes.

2. Data input

The operator enters the start and end microtitre plate numbers, the gridding pattern, the number of replicates in the gridding pattern and the sterilising conditions to be used on the gridding tools. Additionally, the sample identities are entered into the database.

3. Documentation

Printouts of the system error logs and the sample identity databases are at the discretion of the operator.

4. Maintenance schedules

Maintenance is divided into two main types, operator checks and preventive maintenance. The latter is covered by agreements with the main suppliers. The operator checks form part of the Visual Basic program front end. The end-user, following SOPs, runs initialisations and simple positional checks daily. Calibration checks are performed weekly although with experience this may be extended to fortnightly or monthly.

5. Personnel

Initially, key personnel were trained on the system by the developer and with the assistance of the developer produced the SOPs and helped to finalise the operational design.

6. Waste disposal

Essentially no waste materials are produced by the system. Normal laboratory disposal procedures apply.

7 Deciphering Genomes Through Automated Large-scale Sequencing

Lee Rowen, Stephen Lasky and Leroy Hood
Department of Molecular Biotechnology, University of Washington School of Medicine, Seattle, USA

CONTENTS

Deciphering Genomes

◆◆◆◆◆◆ I. INTRODUCTION

In 1989, Heiner and Hunkapiller wrote:

> The advent of automated procedures has already greatly increased the number of bases that can potentially be sequenced per year. Anticipating further developments, researchers have begun to undertake huge sequencing projects such as the Human Genome Initiative. Molecular advances in cloning and mapping will certainly accompany the developing automation. In the longer term, entirely new sequencing technologies may replace existing methods. ... Undoubtedly, an ability to determine even very large sequences quickly and reliably will have a profound impact on our understanding of biological processes. It is also quite possible that further progress in automating all of these procedures will mean that sequencing a gene may become as routine in disease diagnosis as growing up a bacterial culture is today.

These optimistic words marked the dawn of automated fluorescent sequencing, when the first Applied Biosystems 370 Sequencer with a capacity of 16 lanes, capable of generating traces 350 bases in length, had been released commercially for two years, with less than 100 machines sold world-wide. Since then, increased lane capacity, faster run times, longer well-to-read distances, improved enzymes, more sensitive fluorescent dyes and better base-calling algorithms have increased the throughput, length and accuracy of sequence reads generated using automated fluorescent technology. For example, the Perkin-Elmer Applied Biosystems Division (PE-ABD) 377 Sequencer now runs 64 lanes, generating sequence reads of 400 bases in 2 hour runs, and 800 bases in 8 hour runs. Since the late 1980s, advances in automated high-throughput DNA sequencing have come largely through evolutionary changes in the Sanger dideoxy method (Sanger *et al.*, 1977). Alternative technologies, such as mass spectrometry (Glover *et al.*, 1995), scanning tip technologies and sequencing by hybridisation (Arlinghaus *et al.*, 1997; Drmanac *et al.*, 1992, 1993; Strezoska *et al.*, 1991), have not succeeded in matching the throughput or accuracy afforded by automated fluorescent sequencing. Consequently, it is likely that the first reference sequence of the human genome will be largely obtained using variations on the current methodology.

Sequencing has proceeded rapidly since Ansorge's laboratory determined the first contiguous stretch of human genomic sequence greater than 50 kilobases (kb) using automated fluorescent sequencing technology in 1988 (Edwards *et al.*, 1990). Twelve microbial genomes in the 700 kb to 5 Mb size range have been completely sequenced. The yeast genome (12 Mb in 16 chromosomes) has also been completely sequenced (Cherry *et al.*, 1998; Goffeau *et al.*, 1996). For a current listing of published sequences, see the Microbial Database at http://www.tigr.org. The round worm *Caenorhabditis elegans*, with a genome size of 100 Mb, is about 70% complete (Ahringer, 1997; Kuwabara, 1997) and should be finished in 1998. Genomic sequencing of the fruitfly *Drosophila melanogaster* (150 Mb) and the common weed *Arabidopsis thaliana* (100 Mb) (Bevan *et al.*, 1998; Goodman *et al.*, 1995) are under way and should be completed within a few years.

Because the genomes of significant crops such as corn and soybean (Goldberg, 1978; Gurley *et al.*, 1979), and significant mammals such as human and mouse, are more complex and much larger (over 1 gigabase), the sequencing of these genomes presents significant challenges. These challenges include clone acquisition and mapping, extensive automation of the sequencing process, and reconstruction of long stretches of contiguous sequence that faithfully represent the genomes from which they are obtained. A discussion of how these challenges are currently being met is the subject of this chapter.

◆◆◆◆◆◆ II. SEQUENCE SAMPLING OF COMPLEX GENOMES: ESTS, STCS AND REGIONAL CONTIGS

Short of complete genomic sequencing, three distinct high-throughput sequencing approaches have been used to sample the features of complex genomes. These include expressed sequence tags (ESTs) (Adams *et al.*, 1992, 1993; Gerhold and Caskey, 1996), sequence tag connectors (STCs) (Venter *et al.*, 1996) and targeted sequencing of selected regions of the genome (regional contigs) (Kawasaki *et al.*, 1997; Rowen *et al.*, 1996). All three approaches require the construction of clone libraries, preparation of DNA templates from large numbers of clones, sequencing of the DNA templates, and computational processing of the sequence data for downstream analyses and dissemination to the community. Consequently, advances in automated sequencing technology resulting from the implementation of any one of these strategies naturally benefit the others.

A. ESTs

The EST approach, pioneered by Craig Venter and his co-workers (Adams *et al.*, 1992), serves to identify coding sequences (genes). Using plasmid vectors, cDNA libraries are constructed from mRNAs isolated from specific tissues in such a way that the 5′ and 3′ ends of the message can be identified by sequencing one or both vector–insert joins of the plasmid. In contrast to full-length cDNA sequences, ESTs typically give only partial coverage of the coding sequence and 3′ untranslated regions of the genes. Even so, ESTs have proven invaluable for gene identification and for delineating the anatomy of tissue-specific and developmental stage-specific gene expression. More than 1 million human and 200 000 mouse EST sequences are stored in GenBank, thanks to three large-scale sequencing efforts: the Institute for Genomic Research (TIGR), the Merck-St Louis effort, and the Cancer Gene Anatomy Project (CGAP). A database, termed Unigene (http://www.ncbi.nlm.nih.gov/Unigene/index.html), has identified about 50 000 distinct human genes, based on a computational analysis of cDNAs, ESTs and annotated coding sequences in the various genome-related databases.

B. STCs

In contrast to ESTs, which specifically sample transcribed regions, sequence tagged connectors (STCs) sample short stretches of random genomic sequence, most of which is non-coding. The STC approach (Venter *et al.*, 1996) was proposed to facilitate the extensive clone acquisition and mapping required for sequencing the human genome (see discussion below). STCs are generated by sequencing both vector–insert joins (ends) of randomly generated clones obtained from highly redundant libraries made from whole genome or chromosome-specific DNA. Unlike sequence tagged sites (STSs) (Olson *et al.*, 1989), which are unique PCR-based sequence markers in the genome, STCs are directly associated with clones which provide sequence-ready source material for large-scale chromosomal sequencing (see below). The STC approach is currently being applied to clones of human DNA obtained from bacterial artificial chromosome (BAC) libraries that have received the proper human subjects' approval (Report of the Task Force on Genetic Information and Insurance, 1993). The current human genome BAC libraries have an average insert size of ~150 kb. STC sequencing for the human genome is currently under way at TIGR and the University of Washington. By the end of 1999, there should be 600 000 STCs (G. Mahairas, personal communication). Assuming an average STC sequence read length of 500 bases, sequences obtained from both ends of 300 000 BACs (a 15-fold coverage) would collectively cover 10% of the human genome. Assuming a random distribution of STCs across the genome, one would expect an STC to be found on average every 5 kb. STC sequencing is also being used to sample the *Arabidopsis thaliana* genome (at TIGR and several other facilities), and corn and soybean genomes (at the University of Washington). For relatively uncharacterised genomes, the STC approach provides a rapid way to sample the genes (and regulatory regions), and characterise the distinct types and prevalences of genome-wide interspersed repeats.

C. Regional Contigs

In theory, the complete reference sequence obtained for the human genome would contain 24 strings of Gs, Cs, As and Ts, one for each of the 22 autosomes and the X and Y chromosomes. In practice, the human genome sequence is now being determined from selected regions of specific chromosomes for which there are physical maps of sequence-ready clones. As of the beginning of 1998, about 3% of the human genome has been sequenced, with the longest contiguous stretch of sequence being 1.5 Mb (Rowen *et al.*, 1997). Contiguous stretches of genomic sequence of the order of hundreds of kilobases are typically obtained using the random or "shotgun" approach, wherein large insert source clones (e.g. BACs) are randomly fragmented into small (1–3 kb) pieces which are subcloned into a phage or plasmid vector such as M13 or pUC to generate a shotgun library for sequencing. A sufficient number of clones from this library are sequenced to obtain, on average, six- to ten-fold redundant coverage of the original insert. The sequence reads (500–800 bases long) from the shotgun library are assembled into multiple sequence alignments (contigs) by identifying overlaps between the individual reads, and a "consensus" sequence for the original source clone is determined from the best data. Longer sequences are constructed by joining the consensus sequences of overlapping clones.

Regional sequence determination from long stretches of genomic DNA is the most difficult of the three approaches to sampling complex genomes because of the mapping required to produce a minimum overlapping set of clones. None the less, it is the type of sequencing with which our laboratory has gained the most experience. Consequently, it will form the basis for the remainder of our discussion of automated sequencing technology (Table 1).

Table 1. Some regional chromosomal sequences determined in the Hood laboratory

Region	Organism	Chromosome	Length (kb)
T-cell receptor alpha/delta	human	14	1072
T-cell receptor beta	human	7	685
T-cell receptor beta	mouse	6	701
Major histocompatibility, class III	human	6	342
Major histocompatibility, class II–III	mouse	17	710
Total			**3510**

♦♦♦♦♦♦ III. LARGE-SCALE GENOMIC SEQUENCING: FROM CLONE ACQUISITION TO AUTOMATED ANALYSIS

A. Complexity of the Human Genome

The human genome, as well as most other complex eukaryotic genomes, differs from bacterial genomes in four important respects:

1. Eukaryotic genomic DNA is organised into chromosomes with distinctive banding patterns. On a molecular level, bands are characterised by differences in GC content and repetitive DNA content (Benbow, 1992; Holmquist, 1989).
2. Most of the DNA in higher eukaryotic genomes is non-coding. Human DNA, for example, is believed to contain only ~2–4% coding sequence.
3. Complex eukaryotic genomes are highly repetitive (Smit, 1996). For example, the human genome may average 40–45% repeat sequences (Arian Smit, personal communication). These repeats include:
   (a) satellite DNA at the centromeres and telomeres of chromosomes;
   (b) clusters of ribosomal RNA repeats (Hsu *et al.*, 1975);
   (c) genome-wide interspersed repeats (e.g. transposable elements and retroviral skeletons (Jurka *et al.*, 1992; Smit, 1996);
   (d) locus-specific repeats (e.g. T-cell receptor and immunoglobulin gene duplications) (Kawasaki *et al.*, 1997; Rowen *et al.*, 1996); and
   (e) chromosomal duplications and translocations (e.g. olfactory receptors) (Trask *et al.*, 1998).
4. Complex genomes are polymorphic (Charmley *et al.*, 1994; Epplen *et al.*, 1997; Nickerson *et al.*, 1992; Rieder *et al.*, 1998). With the exception of inbred strains (e.g. mice) the sequence of genomes of organisms within a species can vary significantly. In humans, for example, the two sets of chromosomes in an individual will differ, on average, by one nucleotide in 500 to 1000 bases when orthologous sequences are compared for base substitution (Fullerton *et al.*, 1994; Nickerson *et al.*, 1992). Because of differences in the copy number of duplicated regions,

orthologous chromosomes can vary significantly in overall DNA content, up to 40% in some cases (Mefford *et al.*, 1997).

The highly repetitive and polymorphic nature of human DNA causes difficulties for mapping and sequencing the genome. These difficulties have been overcome largely by developing methodologies that utilise redundancy of clone coverage as a check for internal consistency of clone maps and consensus sequences, the specifics of which will be described below.

B. Current NIH Guidelines for Sequencing the Human Genome

Federally funded sequencing of the human genome, scheduled to be complete in 2005, is currently governed (in the United States) by the following guidelines:

- Sequenced clones must be obtained from clone libraries that have the proper human subjects' approval. In practice, this means that the DNA donor must be anonymous.
- Gaps in the consensus sequence of a clone must be filled and/or annotated when they are not filled.
- The overall accuracy of a consensus sequence must exceed 99.99% (i.e. less than one error per 10 kb).
- Genome centres must determine long contiguous stretches of sequence by identifying and sequencing a minimal tiling path of overlapping clones. These clones must faithfully represent the genome.
- Data must be released to the community in a timely manner: sequencing in progress must be posted on web pages and finished sequence must be released to GenBank (Benson *et al.*, 1998).

In most cases, these guidelines have evolved from discussions among the key players in the human genome sequencing effort. As the consensus in the community changes, so do the guidelines.

The US federal agencies' (NIH and DOE) objective is to sequence about 60–70% of the human genome. The remainder will be done by other countries (e.g. England, Germany, France and Japan).

C. A Brief Comment About Sequencing Strategy

Implicit within the guidelines of the funding agencies is the assumption that the cost per finished base must decrease significantly without a sacrifice in data quality. Genome centres whose costs are too high will cease to be competitive and, consequently, will not receive continued funding. Costs can be attributed to equipment, supplies, labour and institutional overheads.

The need to sequence DNA cheaply, efficiently and carefully makes automation of the overall process a desirable goal. Successful automation reduces labour costs, improves the consistency of data quality and increases efficiency and throughput, all of which should in principle drive down costs. At present (early 1998), significant progress in automating high-throughput sequencing (template preparation, performing sequencing reactions, electrophoresis, base-calling) has occurred (Du *et al.*, 1993; Huang *et al.*, 1994; Mardis and Roe, 1989; Wilson, 1993). Not yet automated are the processes of clone acquisition, minimal tiling path construction, and consensus sequence determination ("finishing"). These labour-intensive bottlenecks are a current focus for automation in the genome sequencing community.

The need to drive down costs influences the choice of sequencing strategy. Throughout the 1990s, there has been significant controversy among sequencers regarding the most efficient and cost-effective procedures for large-scale genomic sequence determination. On the one hand, so-called directed sequencing, wherein an unknown sequence is determined from priming off a known sequence using custom oligonucleotides, has a certain conceptual appeal (Voss *et al.*, 1995). This method was employed to sequence 77 kb of the human T-cell receptor beta locus in 1991 (Slightom *et al.*, 1994). In principle, the sequence can be determined by sequencing each DNA strand and resolving the differences. However, this form of directed sequencing has two disadvantages. First, it is slow because one sequence must be obtained before the next round of adjacent sequencing can occur. To make the process efficient for large inserts, several different starting-point sequences would need to be extended simultaneously. Second, interspersed repeats and locus-specific repeats make it difficult to prime uniquely through some regions of the source clone.

A second semi-directed approach, in use at some genome centres, involves the subcloning of a source clone into plasmids, seeding of these plasmids with transposon insertion sequences, mapping the transposon insertions by PCR to determine their relative positions in the clone, and sequencing a minimal tiling path of plasmids by priming off the successively ordered transposon tags (for an example, see http://www-shgc.stanford.edu/Seq/Protocols/BacMat.html). This strategy assumes that the PCR mapping steps required to order the transposon tags are cheaper and more efficient than the shotgun sequencing that they are designed to replace (see below).

A third approach involving a significant mixture of both the shotgun and directed strategies, in use at several genome centres, e.g. the Washington University Genome Center in St Louis, involves the generation of a shotgun library from the source clone in M13 or plasmid vectors, and sequencing of this library to a mid-range of redundancy (e.g. five- to six-fold). At this level of redundancy, numerous gaps exist between sequence contigs. Moreover, there are several regions where only one strand of the source clone is covered by sequence reads and low quality regions of sequence must be improved in order to achieve the current standards of accuracy. Directed methods based on custom oligonucleotides and PCR are used to fill gaps and resolve low quality areas in the sequence. This method places a significant burden on "finishing", which can comprise more than half of the labour and more than 90% of the total required time for obtaining the sequence of a source clone.

The high-redundancy shotgun approach (~eight- to ten-fold), also in use at several genome centres, e.g. University of Washington (Roach, 1995; Rowen and Koop, 1994), has the advantage of reducing the finishing burden but the disadvantage of requiring a large amount of sequencing, which is expensive. Shotgun sequencing, however, is fairly easy to automate and, therefore, can be done efficiently. High-redundancy shotgun sequencing has become significantly more popular over the past two years, partly because of improvements in assembly programs, and partly because of reduction in the number of sequence reads required due to the longer and more accurate reads arising from improved sequencing enzymes and more sensitive fluorescent dyes.

In terms of strategy, new technologies can rapidly change the overall picture of genomic sequencing. Arguments that seem convincing from a given set of assumptions (about, for example, read length, cost of reagents, capacity for automation) can seem antiquated when the assumptions upon which they are based are no longer valid or necessary.

D. An Overview of the Entire Process of Genomic Sequencing Using the High-redundancy Shotgun Method

Genomic sequencing in the context of the Human Genome Project begins with the identification of a target region of the genome to be sequenced and ends with the submission of a consensus sequence with an error rate of less than 1 base per 10 kb to the public database. Genomic sequencing encompasses the following steps:

- Clone acquisition
- Identification of a minimal tiling path
- Clone validation
- Shotgun library construction
- DNA template preparation
- Shotgun sequencing
- Assembly
- Finishing
- Verification of the consensus sequence
- Submission of the sequence to GenBank.

These steps will be described in this subsection. In the section IV, on page 178, we will discuss the current status and challenges of automation with regard to these steps and overall systems integration.

1. Clone acquisition

In the United States, the source clones for sequencing must be acquired from NIH-approved clone libraries. These libraries are typically constructed by using restriction enzymes at low concentration to partially digest genomic DNA isolated from blood, sperm, or cell lines. The partially digested DNA is size-selected by pulse field gel electrophoresis or sucrose gradient ultracentrifugation and subcloned into a vector capable of propagating the human DNA in *E. coli*. In the past, cosmid vectors, containing inserts of 30–45 kb, have been used for library construction. Currently, PAC (P1 artificial chromosomes) or BAC (bacterial artificial chromosomes) vectors capable of propagating larger inserts (~60–250 kb) are being used to make the ethically-approved libraries. PAC and BAC clones are superior to cosmids for three reasons: they appear to be more stable (Shizuya *et al.*, 1992); their insert-to-vector ratio is higher hence, less vector is sequenced in large-scale projects; and fewer clones need to be mapped and sequenced to cover long stretches of the genome. The ideal clone library is highly redundant, containing at least 15 genome equivalents with randomly cleaved inserts to ensure complete representation of the genome among the clones. PAC or BAC clones from the genomic libraries are typically stored as frozen cell stocks in sets of 384-well microtitre plates and numerous replicas are made. The providers of clone libraries generally make available filter sets of arrayed clones suitable for hybridisation screening and/or pools of clones suitable for PCR screening.

To acquire clones for sequencing a target region, genomic libraries must be screened. This is currently done in one of two ways. In the first method, a probe is prepared from a PCR product made from known unique sequence, usually either a cDNA or an STS whose map position in the genome is known. The probe is used to hybridise an arrayed filter set of clones from the library. Alternatively, PCR with primers from known unique sequence is used to identify clones from pools of the library. With appropriate pools, the identity of individual positive clones can be determined. Candidate clones identified by one of these methods are procured from the library and subjected to further validation to ensure that they faithfully represent genomic DNA. To make hybridisation screening efficient, several probes can be used simultaneously in one round of hybridisation. After all of the positive clones have been obtained, they are sorted out using PCR with the primers from which the individual probes were made.

When a reasonably redundant STC resource is available, a third approach to library screening will be possible, namely, computer-assisted identification of STC matches to pre-existing genomic sequence in the region of interest (Venter *et al.*, 1996). For example, the STC resource from a 15-fold redundant BAC library (600 000 STCs, 300 000 fingerprints) will contain, on average, 30 STCs for a 150 kb insert BAC. The BAC clones minimally overlapping the 5′ and 3′ ends of a pre-existing stretch of sequence can then be chosen for further sequencing for the purpose of extending the length of the sequence contig. Clone candidates based on STC hits can be procured from the libraries and subjected to further validation procedures. This process can be repeated until sequence contigs are merged. STC-based library screening will involve considerably less work and expense than either of the two wet-laboratory approaches currently being used. Thus, the STC resource makes any portion of the characterised genome readily accessible to large-scale sequencing.

2. Minimal tiling path construction

In order to sequence 1–2 Mb stretches of genomic DNA efficiently, a minimally overlapping set of source clones (minimal tiling path) must be identified. Two strategies can be distinguished in this regard: the first-map-then-sequence approach, and the first-sequence-then-map approach. Each of these strategies and their tactical variations will be discussed in turn.

(a) First-map-then-sequence: building megabase-sized physical maps

Current practice in the human genome community is for sequencing centres to identify large target regions, typically entire chromosomes or portions thereof. For the sake of discussion, we will focus on minimal tiling path construction for megabase-sized target regions. In the first-map-then-sequence approach, a sufficient number of clones from the target region are acquired from clone libraries and then mapped, that is, ordered along the chromosome with respect to each other. From the physical map, a minimally overlapping set of clones is identified for sequencing. Physical mapping of long contigs has the disadvantage of being slow, but the advantage of allowing genome centres to determine long contiguous stretches of sequence quickly, once the physical maps are in place, because multiple clones in the minimal tiling path can be sequenced simultaneously.

Four tactical variations of physical mapping have been employed by genome centres for sequencing megabase-sized regions:

1 *Subcloning YACs into cosmids.* In the early 1990s, several whole genome libraries were prepared using the yeast artificial chromosome (YAC) vector. Inserts subcloned in YACs are large, typically several hundred kilobases up to 1 Mb in size. Because of the complexity of the human genome and the fact that YACs are intermixed with yeast chromosomes, it is not technically feasible to sequence YACs directly. Instead, YACs mapped to a specific region and verified for clonal fidelity to the genome (i.e. shown to be not chimeric, deleted, or rearranged) are subcloned into cosmid vectors, which contain 30–45 kb inserts. Cosmid clones are picked at high redundancy and mapped by restriction digest fingerprinting (Wong *et al.*, 1997). From the maps, a minimal tiling path of cosmids across the original YAC clone is chosen for sequencing. Longer physical maps are constructed by subcloning overlapping YACs.

   The Olson Genome Center at the University of Washington has implemented this strategy to construct the longest contiguous stretch of human sequence to date – 1.7 Mb (as of February 1998) (http://www.genome.washington.edu/uwgc/). This YAC-based approach must now be discontinued for the US Human Genome Project because the YAC libraries lack the appropriate human subjects' approval.

2 *Screening approved genomic libraries with a dense set of mapped genetic markers.* If the target region has been sufficiently saturated with genetic markers such that the average distance between the markers is less than the average insert size of the source clones (150 kb for BACs), then a set of adjacent markers covering, say, a megabase, can be used simultaneously to screen a library by hybridisation. Assuming randomness

and redundancy of the library, enough clones should be obtained to construct a megabase-sized physical map, from which the minimal tiling path can be determined using restriction digest fingerprint analysis.

3 *Creating a dense set of markers for clone acquisition.* Unfortunately, in the overall scope of the Human Genome Project, densely mapped sets of markers are rare. Instead, the mapped markers may be hundreds of kilobases apart. Current STS maps have an average spacing of 200 kb, which is larger than the average insert size of the source clones, and in many regions the markers may be a megabase or more apart (Hudson *et al.*, 1995). To solve this problem, several genome centres are obtaining YAC clones that cover their regions of interest. Even though YACs are no longer approved as source material for sequencing, they can still be used to generate a set of densely spaced markers suitable for screening. To do this, a YAC is fragmented and subcloned into a plasmid or M13 sequencing vector. From a couple of hundred sequence reads, unique (i.e. non-repetitive) stretches of sequence can be identified and used as probes to screen approved libraries. For this approach to work, YACs must be chosen that are not chimeric (McCormick *et al.*, 1993; Selleri *et al.*, 1992; Wada *et al.*, 1994) and that do not contain a chromosomal duplication (see the discussion on clone validation below).

4 *Cluster mapping.* If the choice is made not to increase the density of markers from a given region, then widely (>200 kb) spaced markers can be used to procure non-overlapping clusters of clones. Vector–insert joining (end) sequences can be obtained from selected clones in each cluster and, if unique, used to make probes for a subsequent round of library screening. Through successive rounds of screens, clusters of clones are merged into megabase-sized clone contigs from which a minimal tiling path for sequencing can be determined using fingerprint analysis.

(b) First-sequence-then-map: the bootstrap approach to minimal tiling paths

Most genome centres currently have a sequencing capacity of 2–30 Mb/year (Rowen *et al.*, 1997). Assuming an average clone insert size of 150 kb, a 30 Mb/year operation needs to sequence 200 BAC clones per year, or 17 BAC clones per month. To meet the Human Genome Project's year 2005 goal, genome centres' sequencing capacity must be significantly increased, possibly to hundreds of megabases/year, depending on the number of funded centres. There are simply not enough physical maps from ethically-approved libraries to come remotely close to supplying enough source clones for sequencing at the required throughput. Therefore, many argue that a map-first-then-sequence strategy is too slow and that a different strategy must be adopted:

1 *Iterative library screening and sequencing.* Using markers from the target region to screen libraries, clusters of clones can be procured as described above. Clones that are internally contained within the cluster, so that genomic validation can be determined by comparison with

overlapping clones, are sequenced. These are called nucleation or seed clones. In the meantime, new probes are identified from the end sequences of clones at the outer edges of the cluster and used to screen libraries. Clones acquired in this manner are end-sequenced, and the end sequences are aligned against the nucleation sequence to determine their map position. Clones with minimal overlap to the nucleation sequence at each end and whose fingerprints are consistent with other clones in the cluster are sequenced. Using this method of contig building, ~140 kb of sequence can be added to each end of the contig with each round of contig extension, assuming BACs with 150 kb inserts are used, and assuming an average of 10 kb of minimal overlap per end. For this approach to be efficient, several contigs must be sequenced simultaneously.

2 *The STC approach: utilising a front-end characterised clone resource.* In 1996, Venter, Smith and Hood (Venter *et al.*, 1996) proposed a strategy aimed at ameliorating the clone acquisition problem for high-throughput sequencing. The so-called STC (sequence tag connector) approach requires that a few hundred thousand BAC clones from ethically approved libraries be characterised in the following manner:

(a) ~400–500 bases of sequence are obtained from each of the two vector–insert joins of each clone (end sequences, or STCs);
(b) a single fingerprint for each clone is obtained with a 6-base cutter restriction enzyme.

The STC and fingerprint data are to be made publicly accessible via the web so that researchers can identify STC matches to pre-existing stretches of genomic sequence. Clones that minimally overlap the sequence can then be ordered from the commercial provider of the clone library from which the STCs are derived. For the STC resource to be useful, rigorous sample-tracking procedures must be employed in the generation of the STC sequence and fingerprint data (see discussion below on laboratory management information systems). Generation of the STC resource is currently underway at TIGR and the University of Washington. By the end of 1999, there should be 600 000 STCs, which will give one STC per 5 kb, assuming a random distribution of clones in the libraries.

Although the cost of generating a centralised STC resource is significant (US $10–12 million), individual genome centres will be able to drive down their mapping costs by taking advantage of the STC resource. Time, labour and expense will be saved by eliminating most of the large numbers of hybridisation screens required for clone acquisition and replacing them with computer screens. Using the STC approach, nucleation sequences are successively extended by sequencing clones identified via STC matches to the predetermined genomic sequence. To make the process efficient, multiple contigs can be extended simultaneously. As sequence contigs merge, new contigs must be seeded in order to maintain an optimal number of growing points.

3. Clone validation

The procedures used by our laboratory and several other sequencing centres to validate clones acquired by one of the procedures described above include: (i) confirmation of the presence of an STS marker to sort clones into bins; (ii) use of multiple restriction enzymes to generate fingerprints that allow the ordering of clones in a cluster and that also allow internal consistency checks among the clones for the purpose of identifying likely chimeras or deletions; and (iii) use of fluorescence *in situ* hybridisation (FISH) to verify chromosomal locations and also to detect possible chimeras or chromosomal duplications. Even if clones are acquired using the STC resource, they still need to be validated for their faithfulness to the genome, according to the current NIH guidelines.

Three interrelated challenges confound the task of clone validation:

- Randomness, redundancy, and fidelity of the libraries.
- Complexity of the experimental procedures.
- Complexity of the genome – duplications and polymorphisms.

(a) Randomness, redundancy and fidelity of libraries

Because PAC and BAC libraries are prepared by using partial restriction enzyme digestion on genomic DNA, there is the potential for non-randomness because of differences in the distribution of restriction sites (e.g. fewer or more, depending on the GC content) and because of the preferential cleavage at some sites. This difficulty can be partially overcome by using more than one enzyme to make the libraries. Non-randomness can also be partially overcome by increasing the redundancy of clone coverage in the libraries. The redundancy of the ethically approved libraries will be eventually 15- to 20-fold or higher. In terms of randomness, there are grounds for optimism. The >20 000 human STCs generated from the ends of >10 000 BACs appeared to be randomly distributed by a variety of criteria, e.g. expected number of EST, genomic sequence, simple sequence repeat and genome-wide repeat hits as well as the observation that no STCs were identical, implying that identical sites were not cleaved repeatedly (G. Mahairas, personal communication). In terms of fidelity and stability, BAC libraries appear to have a good track record (Shizuya *et al.*, 1992). Their major limitation appears to be the occasional integration of bacterial transposons into the human DNA inserts.

(b) Complexity of the experimental procedures

Library screens carried out by hybridisation with a mixture of several probes can suffer from a significant incidence of false positives if: (i) the hybridisation conditions are not optimised; and (ii) one or more of the probes is not unique. Obtaining reproducible and sharp restriction fingerprints of BAC clones depends on variables such as the age of the buffer, the effectiveness of the cooling system and the voltage used for the electrophoresis. Furthermore, BACs vary significantly in their size range. If a relatively infrequent 6-base cutting restriction enzyme is used, fewer but larger fragments are obtained, whereas if a frequent 6-base cutter is used, more and smaller fragments are obtained. In the former case, the long fragments are difficult to size accurately; and in the latter case, one gets multiple fragments of the same band size and loses the smallest fragments because they migrate off the bottom of the gel or do not stain well. Because band sizes are hard to measure accurately, and because there are so many bands, one can be misled by the coincidental matches of bands into believing that clones share restriction fragments and hence overlap when in fact they do not. High redundancy of clone coverage helps, but is not always available. Analysis of the fragments produced by additional enzymes can help to confirm genuine overlaps.

If the first-sequence-then-map approach is used for minimal tiling path construction, one can draw fairly reliable conclusions about overlaps and internal consistency of clones based on fingerprint data. Additional information provided by genomic sequence can confirm the exact size and order of fragments in the regions of potential overlaps.

(c) Complexity of the genome – duplications and polymorphisms

Unless genomic duplications involve long tandem repeats, BACs can usually be assigned to specific chromosomal locations by fluorescence *in situ* hybridisation (FISH), which has a resolution of ~3 Mb. Thus, FISH is quite useful for sorting out clusters of clones that are positive for a given marker but which have distinctly different fingerprint patterns. For example, FISH was used successfully to identify the chromosomal location of a cosmid containing a trypsinogen gene that is 93% similar to the other trypsinogen genes, but which could not be mapped with the other trypsinogen-containing cosmids by fingerprints (Rowen *et al.*, unpublished data). Indeed, it mapped to another chromosome and was part of a large regional chromosomal duplication. If a BAC being considered for sequencing maps to two chromosomal locations by FISH, it could either be chimeric or part of a chromosomal duplication. In the latter case, several of the BACs in a fingerprinted cluster should consistently map by FISH to the same locations.

Polymorphism potentially poses a serious challenge to clone validation. Each of the ethically approved libraries contains two copies of each autosome, which on average will differ from each other by ~0.2%, a polymorphism rate greater than the acceptable error rate for sequencing (0.01%). Unless the variations happen to be RFLPs for the restriction enzymes used for internal consistency checks, most of the nucleotide substitution variations and small insertion–deletion variations will be undetectable by the clone validation methods. With sufficient redundancy of clone coverage, even RFLPs can be sorted out by their fingerprint patterns because multiple clones (not just a single clone) will exhibit distinct restriction patterns for each of the two haplotypes. Medium-size insertion–deletion polymorphisms (e.g. 0.3—50 kb) will generate highly confusing fingerprint patterns at low redundancy of clone coverage. As a result, it may be difficult or impossible to distinguish clearly an artifactual clone from an insertion–deletion polymorphism. Such size polymorphisms may be rather common. Thus, it is important to have sufficient depth in clone coverage so that restriction patterns due to different haplotypes can be distinguished from fingerprint inconsistencies due to the occasional artifactual deletion or insertion.

All of these complexities pose significant problems for validating nucleation BACs, where internal consistency among fingerprint patterns is the primary method of clone validation. For BACs used in contig extension, fingerprints can be compared to the pattern predicted by the pre-existing sequence, which will greatly assist in sorting out the data. It is clear that highly redundant, well characterised, properly constructed clone libraries are crucial.

The foregoing discussion demonstrates that clone validation is not trivial, especially when large insert clones are used for sequencing. We have discussed clone acquisition and validation in considerable detail, even though these procedures are not part of automated sequencing *per se*, because they constitute a significant bottleneck in the process of large-scale genomic sequencing.

Table 2 summarizes strategies commonly used for clone validation.

4. Shotgun library construction

Most genome centres employ the shotgun strategy for generating consensus sequences of their source clones (Rowen and Koop, 1994). Centres

Table 2. Strategies for clone verification

Clone acquisition procedure	Verification protocol	Types of difficulties detected	Probable explanation
"positive" from hybridisation screen	PCR with probe primers	no product	the wrong BAC clone was picked; non-specific hybridisation
"positive" from PCR test	fingerprint with 3 restriction enzymes	bands do not fit with the rest of the cluster	chimeric BAC; deleted BAC; insertion–deletion polymorphism; RFLP
"positive" from STC hit in BAC database	fingerprint with 2 restriction enzymes	fragment sizes disagree with seed sequence	a repeat sequence in the genome; clone tracking problem; also see above
"positive" from fingerprint matches	fluorescence *in situ* hybridisation (FISH)	wrong chromosome or >1 chromosome	STS or seed BAC was wrong; chimera; genomic duplication

differ on the level of redundancy they employ in the shotgun phase, but few use less than five-fold or more than ten-fold. The arguments in favour of high redundancy (~eight-fold) shotgun sequencing are:

- There are few gaps to fill, and gaps are generally small.
- More accurate sequence is produced, because of extensive coverage on both strands. Consequently, the burden of "finishing" is reduced.
- Genome-wide and locus-specific repeats can usually be resolved because of the sequence accuracy.
- Most of the procedures (template preparation, sequencing, base-calling) can be automated, so the production of the shotgun reads is efficient.

Shotgun libraries are generally made by fragmenting DNA from the source clone using one of the following methods: sonication (Deininger, 1983), nebulisation (Hengen, 1997) (http://www.genome.ou.edu/Seqstrategy.html), or high pressure disruption (French press or high pressure liquid chromatography) (for example, see http://sequence-www.stanford.edu/group/techdev/shear.html). After fragmentation, repair of ends, and size selection to eliminate the overly small (< 1 kb) or large (> 3 kb) fragments, the shotgunned DNA is ligated into either an M13 or plasmid vector, transformed into competent cells, and plated out at the desired number of clones.

M13 and plasmid vectors each have advantages and, consequently, both are in common use among genome centres. Sequence reads from the single-stranded M13 vector are generally of high quality, and M13 template preparations are cheap and easy to automate. On the other hand, with double-stranded plasmid vectors, two sequences can be derived from one template without employing PCR, thereby reducing the number of templates that need to be made for the shotgun phase of sequencing. More importantly, sequence reads from plasmids assist in the ordering of contigs at early stages of the shotgun assembly process because the approximate insert length and orientations of paired sequences are known (Roach, 1995; Roach *et al.*, 1995). In addition, because the two reads from a given plasmid are constrained by their size and orientation, data from plasmids often helps to distinguish complex repeats. Finally, some sequences (e.g. small inverted repeats) appear to be more stable in plasmid vectors than their M13 counterparts, resulting in fewer anomalies in the assembly. On the other hand, plasmid data has generally been of lower average quality than M13 data, but this may be changing due to improved methods of plasmid DNA template preparation and carrying out the sequencing reactions.

Four difficulties can compromise the quality of a shotgun library: nonrandomness of the fragmentation; clones lacking or containing only a small insert; *E. coli* contamination from the growth of the source clone; and cross-contamination from other source clones. The latter two difficulties can be avoided by careful preparation of the DNA from the source clone and careful sample transfer techniques. The former two difficulties can usually be avoided by developing a robust procedure for fragmentation and size selection. Because there is a large chance for error in shotgun library construction, and because sequencing is expensive, genome centres usually evaluate their libraries by sequencing a limited number of clones prior to passing the library on to production sequencing.

5. DNA template preparation

With appropriate assumptions about average read length, the desired level of redundancy for the shotgun sequencing phase, and the expected yield of useful clones (i.e. clones that contain insert DNA from the source clone rather than vector, *E. coli*, or no insert), the number of DNA templates that must be made can be calculated from the size of the source clone. In our laboratory, we assume that we need to obtain a total of 20 sequence reads per kilobase of our source clone, assuming a target redundancy of nine-fold (including finishing reads), an average read length of 650 bases, and an average pass rate of 70%. The pass rate is the percentage of sequenced clones that can be used in the final sequence assembly. For a source clone of 180 kb, for example, the estimated number of required templates is 3600 M13 clones and half that number of plasmid clones.

After plating out the shotgun library, plaques or colonies must be picked and grown overnight. The culture volume required depends on the particularities and yield of the DNA template preparation protocol used in the laboratory. There is no universal agreement on the best methods of template preparation, either for M13 or for plasmids. Since most genome centres post their protocols on web pages, laboratories can pick and choose among the options (see Table 3). Factors that enter into the choice of template purification protocol are: cost, purity and convenience. Most methods employ a 96-well format for cell growth and DNA isolation. Because the M13 phage extrudes from the cell, DNA preparations are relatively simple. After growth, cells are removed by centrifugation, phage is concentrated from the culture supernatant by precipitation, and DNA is extracted from the phage using any one of several possible methods (e.g. phenol–chloroform, sodium iodide, or ethanol–butanol). For plasmid DNA purification, the DNA must be extracted from the cell (typically done using alkaline lysis), separated from the *E. coli* chromosomal DNA and proteins (e.g. by deproteinisation and selective precipitation or by using commercially available columns), and treated with RNAase. The AutoGen 740, a robot for DNA purification, does an excellent job in purifying both plasmid and BAC DNAs.

6. Production of shotgun sequencing reads

Over the years, several sequencing chemistries have been developed for automated fluorescent sequencing. Because sequencing is a derivative form of DNA replication, sequencing reactions require a template, a primer, the four deoxyribonucleoside triphosphate (dNTP) building blocks, and a DNA polymerase. In addition, following the Sanger sequencing method (Sanger *et al.*, 1977), dideoxynucleoside triphosphates, one for each base, are used to terminate chain extension. Commercial providers of sequencing kits, e.g. PE-ABD and Amersham,

Table 3. Web addresses of major genome centres

Genome centre	Web address
Baylor College of Medicine	http://gc.bcm.tmc.edu:8088/home.html
CEPH	http://www.cephb.fr/bio/ceph-genethon-map.html
Children's Hospital of Philadelphia	http://www.cis.upenn.edu/~cbil/chr22db/chr22dbhome.html
Columbia University Human Genome Center Project	http://genome1.ccc.columbia.edu/~genome/
Cooperative Human Linkage Center (CHLC)	http://www.chlc.org/
The Eleanor Roosevelt Institute	http://www-eri.uchsc.edu/
Genethon	http://www.genethon.fr/genethon_en.html
Genome Therapeutics Corp.	http://www.cric.com/
Lawrence Berkeley Laboratory	http://www-hgc.lbl.gov/GenomeHome.html
Lawrence Livermore National Lab	http://www-bio.llnl.gov/bbrp/genome/genome.html
Los Alamos National Laboratory	http://www.lanl.gov/
Oak Ridge National Laboratory	http://compbio.ornl.gov/
Roswell Park Cancer Ctr (BAC/PAC Resource Ctr)	http://bacpac.med.buffalo.edu/
Sanger Center	http://www.sanger.ac.uk/
Stanford Human Genome Center	http://www-shgc.stanford.edu/
Stanford Genomic Resources	http://genome-www.stanford.edu/
Targeted Sequencing at the Univ. of Washington	http://chroma.mbt.washington.edu/sequ-www/
The Institute for Genomic Research	http://www.tigr.org
UK MRC HGMP Resource Centre	http://www.hgmp.mrc.ac.uk/homepage.html
UCB Drosophila Genome Center	http://fruitfly.berkeley.edu/
University of Pennsylvania Computational Biology and Informatics Lab	http://www.cbil.upenn.edu/
University of Michigan Medical Center	http://mendel.hgp.med.umich.edu/
University of Oklahoma Advanced Center for Genome Technology	http://dna1.chem.ou.edu/
University of Texas at Southwestern Genome Center	http://gestec.swmed.edu/
University of Utah	http://www-genetics.med.utah.edu/
University of Washington Genome Center	http://chroma.mbt.washington.edu/seq-www/
University of Washington Multimegabase	http://chroma.mbt.washington.edu/msg-www/
University of Wisconsin E. Coli Genome Center	http://www/.genetics.wisc.edu/
Washington University School of Medicine GSC	http://genome.wustl.edu/gsc/
Whitehead Inst. for Biomedical Research and MIT	http://www-genome.wi.mit.edu/
Yale Chromosome 12 Genome Center	http://paella.med.yale.edu/chr12/

have optimised the ratios of dNTPs to ddNTPs to ensure uniform chain termination for a given DNA concentration, fluorescent dye and enzyme.

In the late 1980s, fluorescent DNA sequencing was typically carried out using either the Klenow fragment of *E. coli* DNA polymerase I (Klenow and Henningsen, 1970) or a modified form of the phage T7 DNA polymerase, called "Sequenase" (Tabor and Richardson, 1987). When "cycle sequencing" was introduced, Taq polymerase (Amplitaq) generally replaced Sequenase as the enzyme of choice for large-scale sequencing, both because the reactions could be automated (i.e. in a thermocycler) and because less DNA template is required for the sequencing reaction, thereby making 96-well format DNA isolation procedures more feasible for routine production sequencing (Civitello *et al.*, 1992; Heiner and Hunkapiller, 1989). Moreover, Taq polymerase and cycle sequencing work better than the Sequenase protocol for double-stranded DNA. In 1995, Stan Tabor discovered, through a series of elegant amino acid substitution experiments, that a substitution of tyrosine for phenylalanine in the active site of Taq polymerase resulted in an increased affinity of the enzyme for ddNTPs (Reeve and Fuller, 1995; Tabor and Richardson, 1995). This modified version of Taq polymerase (ThermoSequenase [Amersham] or TaqFS [PE-ABD]) gives more uniform peaks in the sequencing chromatogram, with a concomitant improvement in data quality.

Over the past several years, the fluorescent dyes used in the sequencing reactions have improved as well. For example, the sensitivity of the fluorescent dyes has been further enhanced by exploiting energy transfer to optimise the absorption and emission properties of the dyes (Hung *et al.*, 1997; Ju *et al.*, 1996a, 1996b; Lee *et al.*, 1997; Rosenblum *et al.*, 1997). Fluorescent dyes can be attached either to the primer or to the dideoxy terminator. When primer-conjugated dyes are used, four sequencing reactions must be done for each template, one incorporating each of the four unlabelled ddNTPs for chain termination. On the other hand, when the dye is attached to the ddNTP terminator, only one sequencing reaction per template is required. Thus, dye terminators offer an advantage in terms of throughput. Moreover, with dye terminators, any unlabelled primer can be used in the sequencing reactions, thus allowing directed sequencing with custom oligonucleotides. Dye terminators also offer an advantage in terms of resolving "compressions", where secondary structure in the molecules results in the appearance of collapsed peaks. Introduction of Tabor's mutation into Taq polymerase has reduced the amount of dye terminator needed for incorporation in a sequencing reaction, with the result that the reactions are less noisy due to contamination from unremoved excess dye terminators. For all these reasons, dye terminator chemistry is increasingly becoming the predominant choice among large-scale sequencing centres.

After the sequencing reactions are performed, polyacrylamide gel electrophoresis is used to resolve the mixture of terminated molecules into a sequencing ladder, with single base resolution over a range of about 10 to 900 bases. Most large-scale sequencing centres currently use the Perkin-Elmer Applied Biosystems Division sequencer (373A or 377) for large scale sequencing. This instrument and its associated technology grew from initial work of the Hood laboratory at the California Institute of Technology (Hood *et al.*, 1987; Smith *et al.*, 1986; Smith, 1989). Other sequencers in use at some sites include commercial instruments marketed by LiCor, Pharmacia and Hatachi. The fluorescent dyes used in the sequencing reactions are excited by a scanning laser and detected by a photomultiplier tube or CCD camera as the electrophoresis proceeds. After acquiring the signal provided by excitation of the dyes, the data is processed by a "base-calling" program, which translates raw signal into a sequence of A, C, G and T bases.

Significant improvements have been made in the throughput, length and quality of sequence reads. Throughput has improved by increasing lane density (16 to 24 to 36 to 48 to 64 on the PE-ABD sequencer) and by decreasing gel thickness (0.4 to 0.2 mm on the PE-ABD sequencer), which facilitates faster runs. Read length has been increased by modifying the sequencer to increase the distance between the well in which the sample is loaded and the site at which the gel is scanned for migration of the fluorescent dyes. Data quality has been improved largely by the modifications in the sequencing chemistries described above. With current techniques, clearly resolved reads of 800 bases are not uncommon, assuming high quality DNA template is used.

There is a trade-off between read length and sequencer run time, especially with the PE-ABD 377 sequencer. For the shotgun phase of genomic sequencing, many genome centres have opted for shorter reads (~500 bases) because of the need to run the machines several times a day to meet throughput goals. On the other hand, with shorter reads, more templates must be prepared, sequenced and electrophoresed in order to achieve a given target of redundancy thus driving up the cost of the shotgun sequencing. Aside from reducing the total number of reads required to sequence a source clone, longer reads also facilitate the assembly of the shotgun reads into contiguous sequence. Until recently shorter reads have been the norm for most sequencing centres, but the tide may be shifting in favour of longer reads because of their cost effectiveness. We estimate that the combined labour and reagent costs for DNA purification and sequencing amount to US$20 per read. Reducing the number of reads per kilobase required to sequence a 150 kb BAC from 30 to 20 would result in a US$30 000 savings on labour and reagents. These savings could be applied to the purchase of additional sequencers for the purpose of increasing the throughput.

7. Assembly

After base-calling, shotgun sequencing reads are transferred to a project directory on a computer where they are "assembled" by one of the several assembly engines in use at various genome centres (Bonfield *et al.*, 1995; Dear and Staden, 1991; Lawrence *et al.*, 1994; Miller and Powell, 1994; Parker, 1997; Parsons, 1995; Swindell and Plasterer, 1997; Green: http://www.genome.washington.edu/uwgc/tools/phrap.html).

The output of an assembly program, viewed in an editor, is a set of "contig layouts" in which overlapping sequence reads, the consensus sequence derived from these reads, and the orientation of the individual reads with respect to the consensus sequence are presented. Most editors for assembly programs allow the trace data supporting a sequence read to be viewed for the purpose of data quality and error-checking.

Assembly engines generally work in one of two ways. First, sequences are added to the assembly one at a time. If an incoming sequence matches the consensus sequence of a pre-existing contig by a specified percentage (e.g. 85%) over a specific length (e.g. 50 bases) that sequence is added to the contig and a new consensus sequence is derived. If a sequence matches the consensus sequence of two contigs such that the sequence bridges the contigs, then the two contigs are merged and a new consensus sequence for that contig is generated. Second, pairwise alignments are done between all possible pairs of sequences in the assembly, and the best matches (in terms of sequence similarity and length) are determined and stored in memory. After all of the alignments are done, the contig layout is constructed as a multiple sequence alignment of the best overlaps among the sequence reads. The DNASTAR Seqman program (Swindell and Plasterer, 1997) is an example of the former type of assembler and the Phrap assembler is an example of the latter.

The Phrap assembler (http://www.genome.washington.edu/uwgc/tools/phrap.html), developed by Phil Green at the University of Washington, is increasingly being adopted by large-scale sequencing centres that use the shotgun approach. Phrap takes advantage of two sets of quality measures in its determination of a consensus sequence from a set of shotgun reads. The first set of quality measures, assigned by the base-caller Phred, pertains to the input sequence data (assessed by peak spacing and signal-to-noise ratios in the chromatograms). Green has correlated Phred quality measures with error probability, based on the statistical analyses of several thousand reads in characterised cosmid data sets for which the consensus sequence is known. The second set of quality measures, assigned by the assembler Phrap, pertains to the amount of supporting data for a given base in the sequence provided by additional reads in the assembly. Higher weight is given to confirmation of a base in the sequence by additional reads from the opposite strand, although confirming reads from the same strand are given some weight. Conflicts between high quality bases in the original trace data (assigned by Phred) decrease the Phrap-assigned quality measure for that base in the assembled sequence. To construct a "consensus" sequence from a set of shotgun reads, the best data are used, based on both the Phred and Phrap quality measures. This approach stands in contrast to the "majority rule" method of deriving a consensus sequence used by other assembly engines.

The output of Phrap is viewed in an editor called Consed (Gordon *et al.*; http://www.genome.washington.edu/uwgc/tools/consed.html), which was designed specifically to take advantage of the quality measures provided by Phrap. Consed uses gradations of colour (shades of grey, white being best) to indicate the quality of individual bases. Colour is also used to indicate which base among the set of reads was used to define the consensus (yellow), which bases in the other reads agree with the base used for the consensus (blue) and which bases in the other reads disagree with the consensus (orange). Consed also has additional features that allow the user to view trace data, to identify regions of low quality in the consensus, and to pick oligonucleotide sequences for directed sequencing to improve problematic regions.

8. Finishing: gap-filling and conflict resolution

Even at high redundancy, shotgun sequencing usually fails to produce enough data to determine a consensus sequence at the required standard of accuracy now being adopted for the genome project (< 1 error per 10 kb). Problems generally fall into one of the following areas:

- Gaps
- Conflicts among the reads
- Mis-assemblies.

Because the inserts of shotgun libraries are not cleaved in a perfectly random manner, and because some sequences reduce the ability of clones to propagate in *E. coli*, selective biases can be noted in the distribution of reads. Moreover, because some sequences are unstable and thereby delete themselves from the cloned inserts, the sequence reads from a shotgun project typically assemble into more than one contig. To join these contigs, additional data must be acquired, either by extending the length of reads by custom primer-directed sequencing of individual subclones or the source clone, by sequencing PCR products designed to cover the gaps, or by subcloning restriction fragments of the source clone and sequencing them. The same strategies must be employed to augment stretches of low quality sequence that occur, for example, when all of the data is derived from the ends of sequence reads, where the base-calling is apt to be erroneous.

Conflicts among the reads in a data set must be resolved in order to obtain a reliable consensus sequence. Conflicts are usually due to one of the following causes: compression of the reads on one strand; discrepancies in the number of bases in a polynucleotide tract (e.g. the poly T tails of Alu repeats); noisy data (e.g. missing bases due to low signal in dye terminator reads or discrepancies in base-calling for microsatellite repeats); selected deletions in a subset of clones covering a given region; or collapsed sequence repeats (see below). In addition to observed conflicts, potential errors (e.g. unresolved compressions) can occur in the consensus sequence if data from only one strand of the source clone has been obtained. To solve most conflicts and to confirm regions of single-stranded coverage, selected subclones from the shotgun library are re-sequenced with an alternative sequencing chemistry that gives a different profile of systematic errors. For example, if most of the shotgun data was obtained using the fluorescent dye primer chemistry, sequencing with dye terminators can be used to resolve compressions. Conversely, if most of the sequencing was done with terminators, sequencing with labelled dye primers can resolve noisy data due to base drop-outs. Because the different chemistries produce different sorts of systematic errors, some genome centres use a mixture of chemistries in the shotgun sequencing phase in order to reduce the number of finishing reads required for conflict resolution (Koop *et al.*, 1993).

Occasionally, a set of shotgun reads will not assemble correctly. Misassemblies can be diagnosed by one of the following methods: detecting a discrepancy between the length of the consensus sequence and the size of the insert in the source clone, judged by fingerprinting; detecting a discrepancy between the predicted fingerprint pattern based on the sequence and the actual fingerprint pattern obtained for the source clone; detecting a "false join" between two contigs; and detecting systematic conflicts between high quality bases when the assembly output is viewed in an editor. Genome-wide interspersed repeats (e.g. LINEs and Alus) and locus-specific repeats (e.g. multiple copies of genes in a source clone) can cause problems for assemblies, especially if the repeats are long and of high sequence similarity (>90% for most assemblers; >98% for Phrap). Redundant data, high quality sequence reads, long reads and data from both ends of plasmids help to resolve difficulties with shotgun assemblies.

Because of the various problems that can occur, finishing usually requires several rounds of additional sequencing before all of the gaps are filled and conflicts or low quality regions are resolved. Most genome centres have a team of "finishers", who are trained to recognise problems and devise solutions. Because finishing is labour intensive, it currently constitutes a bottleneck for high-throughput sequencing.

9. Validation of the consensus sequence

Prior to submission, the consensus sequence for a source clone is typically validated in two ways: comparison of the predicted fingerprint pattern with the actual fingerprint pattern (usually with two enzymes) and comparison of the consensus sequence with sequences obtained from overlapping source clones. In the latter case, conflicts need to be attributed to polymorphisms, rather than sequencing errors, by examining the input sequence data for each clone to verify that the data are of high quality.

Over the past year, the NIH has engaged its genome centres in a "quality control" exercise. For each centre, a limited number of source clones for which sequence has been deposited into GenBank are arbitrarily chosen for evaluation by two other genome centres. The evaluating centres each receive a glycerol stock of the source clone and all of the sequence data that were obtained for the purpose of determining the consensus sequence. The evaluating centres determine, from an examination of these data and additional sequencing done in-house, whether the submitted sequence has met the quality standards set by the NIH. In years to come, this function will be taken over by a "quality evaluation" centre, whose mission will be to monitor the quality of the data submitted by the various genome centres.

10. Data submission

Current practice among genome centres is to release data from assemblies-in-progress on web pages. In some cases, unfinished data are also submitted directly to GenBank. Finished consensus sequences are submitted to GenBank, often with minimal annotation (e.g. clone name, clone library source, chromosome location, interspersed genome-wide repeats). The National Center for Biotechnology Information (NCBI) provides a tool for annotating and submitting sequences called Sequin (http://www.ncbi:nlm.nih.gov/sequin/index.html). Many genome centres develop tools in house that accomplish the same purpose. Some genome centres provide additional annotation, for example sequence variations among overlapping sequences, EST matches, known gene locations and predicted gene locations. After the original submission, GenBank entries can be updated with additional annotation. Because detailed annotation is time-consuming, some genome centres are only annotating features of the sequence that can be identified automatically by database searches.

◆◆◆◆◆◆ IV. SYSTEMS INTEGRATION, AUTOMATION AND TECHNOLOGY DEVELOPMENT FOR HIGH-THROUGHPUT SEQUENCING

Because large-scale genomic sequencing requires a complex series of steps, and because some of these steps are relatively slow (e.g. mapping and finishing), many source clones must be processed simultaneously in a high-throughput operation. A 30 Mb/year operation must complete of the order of four 150 kb BACs per week to meet the throughput requirements. But because the cycle time from clone acquisition to data submission is currently several weeks at best, a sequencing operation must keep track of data pertaining to scores of source clones simultaneously. In order to accomplish this, most genome centres have installed some type of laboratory information management system (LIMS) to facilitate sample tracking, quality assessment, success rates and the like (Hunkapiller and Hood, 1991; Smith *et al.*, 1997) (see the web sites of various genome centres in Table 3 for examples). Well-constructed laboratory information management systems can enable the managers of a sequencing operation to monitor productivity on a quantitative rather than an anecdotal basis. Quantitative data about how the operation is actually working facilitates more rational decisions about how to improve its effectiveness.

Because of the need continually to increase throughput, genome centres typically have a research and development team to evaluate and implement new overall strategies and technical procedures aimed at improving the efficiency of the operation. In the context of production, care must be taken to achieve a proper balance between stability and innovation. On the one hand, entrenchment of procedures that have a proven track record is potentially deleterious. Entrenchment engenders a sociological resistance to change. On the other hand, continuous introduction of new procedures with the aim of improving productivity is potentially destabilising. Constant change in a high throughput operation is more likely to decrease, rather than increase, productivity. To strike the proper balance, choices must be made regarding the adoption of new procedures with an eye to the overall effect on the productivity of the operation. The following questions address the challenges of systems integration in this regard:

- Is the new procedure (machine, protocol, strategy) genuinely better?
- Will the new procedure have hidden or unanticipated adverse consequences?
- Does adoption of the new procedure make sense in terms of other likely developments in the field, current or future?
- Can the new procedure be implemented effectively in the context of the overall operation?

A new strategy or procedure might be deemed an improvement if: (i) it decreases cost; (ii) it increases throughput; (iii) it improves data quality; or (iv) it decreases the cycle time. Ideally, a "better" procedure would do all of these. In practice, there are usually trade-offs or "apples-and-oranges" comparisons. For example, a robot that prepares DNA templates and sets up sequencing reactions automatically might be seen to be advantageous because it would reduce labour costs and increase throughput. On the other hand, if the failure rate of such a machine were 30%, as contrasted to a 10% failure rate of the procedures the machine was designed to replace, the trade-off might not in fact be advantageous, because of the adverse effects of poor data quality on the downstream steps of assembly and finishing.

In this section, we discuss some of the fundamental issues that need to be addressed and solved to build a scalable, high-throughput genomic sequencing operation. These include the following:

- Implementing a sophisticated laboratory information management system (LIMS).
- Designing a production line operation that takes full advantage of the LIMS and best practice strategies and methods moving towards automation.
- Identifying and removing rate-limiting steps and causes of failure in the overall process.
- Integrating the entire system, through a well-developed set of computational tools.
- Disseminating information to the community.
- Hiring effective personnel, and training the technical and managerial staff in the philosophy of the organisation, LIMS and best practice techniques.
- Developing and/or refining emergent new technologies.
- Retooling the production line operation to incorporate new technologies and procedures.

A. Optimising the Overall Operation: The Need for LIMS

Most large-scale sequencing groups have some form of a LIMS to keep track of clones, data pertaining to clones, progress statistics, data quality, pass rates, machine utilisation and the like. These systems range in sophistication from laboratory notebooks and desktop computers at the low end, to database servers with numerous scripts which automatically generate status and quality reports at the high end. The ideal LIMS includes even higher levels of sophistication likely to be essential to the success of a significant scale-up of genomic sequencing. Components of such a LIMS would include:

- *Sample tracking.* All operations performed on a sample or set of samples (e.g. in a 96-well plate) would be recorded and collated using sample tracking barcodes.
- *Troubleshooting/alert.* Technicians would be informed (via hand-held computer or pager) if a process is failing.
- *Enforces consistency.* Because technicians would be logging in and recording all protocols performed on a set of samples, they would be mindful of the need for consistent execution of procedures, resulting in overall higher quality data. Moreover, work performed by individual technicians could be monitored for quality and corrective measures taken as appropriate.
- *Automatic data handling.* Data pertaining to individual samples (e.g. sequence traces) would be sorted and processed in the proper ordering of steps.
- *Report generation.* Data pertaining to samples, projects, machines, and protocols would be queried and sorted in various ways. Plots indicating trends over time would be generated. This would enormously facilitate the management of the sequencing production line.
- *Simulation.* The effect of changes in the operation (e.g. longer or shorter average read length) could be simulated, given the appropriate inputs. This, combined with an assessment of data quality, failure rates and trends, would facilitate sensible decisions of resource allocation aimed at removing bottlenecks and procedure optimisation.

A sophisticated LIMS, such as one currently being implemented in our sequencing centre by Cimarron Software Inc., in Utah (Sargent *et al.*, 1996), has the potential to increase the efficiency of high throughput sequencing in several ways:

- *Early identification of failure.* It is important immediately to flag processes that are failing, for example, degradation in the data quality produced by a particular machine (e.g. thermocycler or sequencer), in the performance of a technician, or in the quality of data produced by a protocol under development. The LIMS would be designed to alert a manager and route data to the appropriate human interpreter when data fails to meet the specifications or quality standards for the process involved. That is, the LIMS would automatically record the execution of processes and indicate readiness for subsequent operations on samples or data *unless* the process or data fails some specification or quality assessment.
- *Data capture.* The proper use of a LIMS will prevent data loss, thus, potentially raising the overall pass rate for sequencing.
- *Bottleneck identification.* A clear understanding of the overall operation provided by the appropriate queries of the database, combined with the simulator, would assist the task of resource allocation to remove bottlenecks.

- *Stimulates a push towards automation.* Procedures would be chosen and developed partly in accordance with their coherence with the LIMS, leading to an overall well-integrated sequencing process that has the potential for scale.

Using a sophisticated LIMS requires a shift in thinking from a research-orientated environment to a factory-style environment, and personnel need to be trained accordingly. Managers must use the LIMS to organise the workflow of their staff, monitor quality and failure rates, and deal with the exceptions, breakdowns and problems that get routed to them for interpretation. Technicians must enter the operations and the data, via barcodes and computers. This requires continued interaction with the LIMS.

B. Automation

In the past ten years, numerous advances in the automation of sequencing have occurred (Adams *et al.*, 1994). Perhaps the two most striking triumphs of automation, as judged by their wholesale adoption by the genome community, are cycle sequencing and computerised base-calling in a fluorescent DNA sequencer. Success here is gauged by the following criteria: (i) efficiency (the procedures are fast and require little or no human intervention or attention); (ii) pass rate (the procedures work on most samples and work most of the time); and (iii) quality (the procedures consistently and reliably produce good data). Well on its way to being universally adopted by the genome community is Phil Green's base-caller, Phred and assembler, Phrap, which automatically generate sequence contigs from sets of high redundancy shotgun data.

Noteworthy advances in automation have been achieved by numerous sequencing groups in the area of building or adapting machines to pick clones, prepare DNA templates, set up sequencing reactions and load gels (for an example of "state-of-the-art" technology development, visit the web site of the Stanford DNA Sequence and Technology Center at http://sequence-www.stanford.edu/). Because of differences in sequencing, and in part because of real or perceived differences in effectiveness and reliability, no robots for clone picking, template preparations, sequencing reaction set-ups, or gel loading have been universally adopted. Determining the best procedures and equipment for high-throughput process automation, therefore, remains an important area of research and development.

In the context of a sophisticated LIMS, automation is desirable because it reduces the number of steps that require human interaction with the database and utilises the LIMS's ability to monitor potential points of failure. Thus, when there is a choice among robots or automated procedures for any given step of the sequencing operation, those that can be built into the LIMS are the preferable options. For example, a machine such as the Packard Multiprobe robot can be controlled remotely using computers. In the ideal production line, there would be a series of machines and procedures that would operate in sequence, automatically recording data into the LIMS, and automatically transferring samples to the next stage in the process.

Procedures where a large number of samples are subjected to a limited number of operations are, by their very nature, a natural focus for automation. In contrast, two procedures essential to sequencing – shotgun library construction and finishing – have generally not been automated or, at best, have been only partially automated. These processes present a challenge to a LIMS-orientated production operation and are a productive area for research.

C. Identifying Rate-limiting Steps and Points of Failure

Assuming that the significant bottlenecks of mapping and finishing can be eliminated, new bottlenecks are likely to appear. These new bottlenecks are likely to be in the production line itself. Moreover, steps where the failure rate is unacceptably high must be identified so that corrective actions can be taken. A LIMS offers two potential advantages in this regard. One is its data flow and process monitoring capacity, which can be used to measure the total time elapsed for sets of processes, the number of samples on which processes are performed, the quality and pass rate of all of the processes, and the changes in these values over time. The second advantage of a sophisticated LIMS that analyses the performance of a strategy is its modelling and simulator functions. Various processes could be provided to the simulator, along with information regarding the number of machines available for each process, the time each process takes, and so forth. Based on these data, one could pose questions such as: If we add another person or thermocycler or sequencer, would the picture change? If the pass rate of a given step were raised by 5%, how would the picture change? Therefore, the use of process monitoring and simulation capacities of the LIMS offers managers objective data regarding the rate-limiting steps and failure points of the operation. With objective data, managers are better positioned to make rational decisions about resource allocation.

D. Informatics: Systems Integration and Data Dissemination

The web, along with platform-independent software applications, has greatly facilitated both in-house informatics (LIMS) and dissemination of data to the community. Solutions are largely in place in the genome community for most of the informatics issues regarding data storage and transfer. Generally, each genome centre develops procedures specific to its own needs and imports those tools that are of general use in the community (Butler, 1994). We cannot over-emphasise the importance of the world wide web as a source of useful tools and data for analysis.

E. The Hiring and Training of Personnel

One process not amenable to automation, yet a central ingredient to a large-scale sequencing operation, is the hiring, training and retention of capable personnel. Many sequencing centres face the challenge of converting an academic research group into an industrial-style operation without incentives such as high salaries and stock options. There tends to be significant turnover of staff, especially at the technician level. In addition, staff scientists at the higher level of management are usually not trained as managers. Their training in personnel management is largely "learn by doing". As groups get larger, more layers of middle management need to be implemented in order to keep subgroups effective and focused. Troubleshooting, both technical and human resource, is continually required.

Effectively managing, training and mobilising groups of people poses a serious challenge for large-scale sequencing. While increasing automation in the sequencing process will reduce human labour, it cannot eliminate it. Once a significant level of scale-up has been achieved, professional managers must be hired to ensure that the operation runs smoothly. A key to effective training is an introductory course that gives both an overview of large-scale sequencing and automation and practical hands-on training. This course sets the tone and frameworks within which staff can grow and mature.

F. Testing Emergent Technologies

Genome centres must pick and choose among the options available for improving throughput. These are likely to include oligonucleotide synthesisers, colony or plaque pickers, robots for DNA template preparation, sample arrayers, sequencing robots, gel loaders and sequence assembly engines. Experience gained by other genome centres is immensely helpful in this regard. Even though any one centre can engage in only a limited number of collaborations with commercial or academic developers, genome centres can collectively explore the terrain and improve prototypes of the new technologies by employing their in-house development resources.

G. Retooling the Operation to Incorporate Changes

The changes that will occur in the genome community over the remaining seven years of the Human Genome Project cannot be fully predicted. There may be advances in sequencing technology that will require an adaptation or overhaul of production facilities. Standards of consensus sequence quality currently endorsed by the community and funding agencies may change. Therefore, a high-throughput sequencing operation must retain a certain level of technical and managerial flexibility. This will require the employment of experienced personnel capable of engineering changes as required. In this regard, the true advantage of a modular design for the sequencing pipeline, and a LIMS that assumes modularity as a premiss, is that upgrading modules at any stage can take place without adversely affecting productivity.

◆◆◆◆◆◆ V. STRATEGIES FOR THE FUTURE

A. The Problem

The sequence of the human genome has been promised by the year 2005. As of today (March 1998), about 3% of the human genome has been sequenced (~90 Mb). In the United States today, there are potentially four or five groups that could, in the next year, scale up to 15 Mb per year. In the rest of the world, there are a few additional groups with that capacity. Two points are key. First, new sequencing centres must be brought to a competitive throughput level. Second, there must be an appropriate balance between resources spent on technology development aimed at increasing sequencing efficiency and decreasing cost, and those spent on production sequencing using today's technology. Unless throughput can be increased by four- to five-fold over the next few years, finishing the human genome at the current standards of quality by year 2005 entails a formidable, if not impossible, challenge.

B. Opportunities to Increase Efficiency of Sequencing

1. Capillary sequencers

Capillary sequencers potentially offer an attractive opportunity to increase sequencing throughput while significantly decreasing the cost and cycle time. Several groups are working on a 96-capillary instrument. The capillary sequencers use perhaps 20–30% of the sequencing reagents and require smaller DNA samples; thus, the costs of reagents for sequencing and DNA purification could be greatly reduced. The gels in the capillary sequencers are pumpable (e.g. monomeric acrylamide) and, accordingly, the process of creating new gels can be completely automated, relieving an enormous bottleneck in current sequencing strategies. Sample loading can also be automated; thus, the entire sequencing process can be run on a 24 hour basis. The cycle time for the capillary sequencers is 1½–2 hours; hence, 12–18 sets of samples could be run during a single day. Realisation of the potential of capillary electrophoresis in a sequence production context would, therefore, have a dramatic effect on throughput. One limitation of the currently available capillary sequencers is that only 450–600 bp of sequence can be generated per capillary. With subsequent development, however, read length may increase.

2. Microfabrication techniques

On the more distant horizon, microfabrication techniques (microfluidics/microelectronics) will create the possibility of developing miniaturised sequencers, integrating together many of the sequencing steps (i.e. DNA purification, PCR amplification, cycle sequencing, sample loading and electrophoresis), and increasing throughput by a high degree of parallelisation (e.g. 1000 or even 10 000 electrophoresis channels). These efforts are now in the very early stages.

3. Increasing sequencing efficiency by increasing acceptable error rate

The currently acceptable standard for the error rate in sequencing is 1/10 000. There are compelling reasons to suggest that a much lower error rate (1/1000) would essentially provide all of the desired information (genes, control regions, etc.) and significantly increase the efficiency and reduce the cost of sequencing. First, the rate of human polymorphisms is about 1/500 to 1/1000. If, for example, a sequence variation is observed between two overlapping clones, without further investigation it is impossible to say whether that difference is an error or a naturally occurring polymorphism. Hence, to have an error rate 10–20 times lower than the polymorphism rate makes little sense. Stringent quality requirements for the human genome sequencing have been justified on the grounds that biologists should have to spend little effort correcting errors. But, in fact, they will still have to investigate every variation in relevant regions to distinguish error from polymorphism. Moreover, emerging DNA chip technology will make the investigation of sequence variations simpler and less expensive than sequencing is today. Second, most of the errors fall in tracks of repeat sequences (i.e. microsatellites or the poly A tracks of Alu sequences), which lie outside coding regions. These are regions that do not encode significant biological information and, accordingly, much higher error rates could be tolerated without loss of the utility of the data. Rather than ensuring that every base in the consensus sequence be of high quality, genome centres could instead annotate stretches of low quality sequence. Finally, real economy in finishing the Human Genome Project could be achieved by reducing the acceptable error rate (so long as the goal of obtaining contiguous sequence is retained). Some of the larger sequencing centres have as many finishers as production line sequencers. The ratio of finishers to production sequencers could be significantly reduced if a lower error rate were acceptable. Moreover, the use of reagents and sequencers could also be reduced.

C. Sequencing Other Complex Genomes

If significant improvements in the efficiency of sequencing come to pass, the technology can be applied to other large and complex genomes such as mouse, corn, rice and soybean. The idea is to make sequencing a complex genome as approachable in the future as sequencing a microbial genome is today. Since in the near future the bulk of the resources will be devoted to finishing the Human Genome Project, the sequencing of these other important genomes must rely upon two large-scale sequence sampling strategies: EST sequencing and low pass genomic sequencing.

1. ESTs

ESTs have proven enormously useful in delineating the gene content of organisms and the expression patterns of genes in various cells and tissues.

Several hundred thousand EST sequences would reveal most of the abundant messages for an organism and many of the rare messages as well. However, EST sampling will miss a significant percentage of the genes: those expressed at very low levels or only for short times in the life cycle of the organism. Moreover, EST data do not reveal many features of the genome: gene family organisation, regulatory regions, genome-wide repeat sequences, syntenic relationships and genetic and evolutionary changes in chromosomes. For insights into these features of the genome, one must employ genomic sequencing.

2. Low pass genomic sequencing

Low pass sequencing can be used in large or small genomes to mine 95–98% of the information for 10% (or less) of the current cost. The idea is to create a 15-fold coverage STC resource for the genome to be analysed (e.g. mouse). Then, one would start sequencing from many points by randomly choosing nucleation BACs. Each BAC insert would be sequenced to a two-fold coverage by the shotgun approach. On average, the Poisson distribution suggests this would provide 85% of the sequence in a multiplicity of contigs. If the double-ended sequencing of plasmids is employed (Roach *et al.*, 1995), then a complete scaffold of plasmid clones (e.g. the linkage of all contigs) can be generated for most BAC inserts. Thus, a particular coding region or regulatory element of interest could be finished, if so desired, by primer-directed sequencing of the relevant plasmid(s). Using STC hits to the sequence scaffold provided by the low pass shotgun of the nucleation BACs, new BACs can be chosen for contig extension. There are two significant areas of cost savings with the low pass approach: (i) in a two-fold vs. eight-fold shotgun project, only 25% as much sequencing is done; (ii) no finishing would be done, saving an additional 40–60% of the labour costs.

By applying this low pass approach, we believe that more than 95% of the genes could be identified and the organisation of gene families determined on chromosomes. This approach will work especially well for mouse, where detailed sequence comparisons with syntenic regions of the human genome will be useful in identifying genes, gene families and regulatory regions. Conserved blocks of sequence will be used to facilitate such analyses. In 10 to 15 years, when we have novel, cheaper, high-throughput sequencing technology, genomes characterised by low pass sequencing can easily be done more accurately and completely. A characterised clone resource obtained using the STC approach will facilitate the easy acquisition of clones for this purpose.

The various EST projects carried out during the 1990s have demonstrated the enormous utility of providing biologists with sequence data as soon as possible. With low pass sequencing, data on important genomes such as mouse could be generated and released quickly to the community, thus facilitating research that will complement the Human Genome Project.

◆◆◆◆◆◆ VI. SUMMARY

Genome centres face the daunting challenge of converting small to medium-scale sequencing operations into highly automated factory-style operations that are capable of processing thousands to hundreds of thousands of samples daily through a series of several processes, many of which currently require a high level of human involvement or intervention. Sophisticated laboratory information management systems, implementation of effective new technologies and strategies, and the recruitment and training of capable managerial and technical personnel are essential to the success of the overall effort.

References

Adams, M. D., Dubnick, M., Kerlavage, A. R., Moreno, R., Kelley, J. M., Utterback, T. R., Nagle, J. W., Fields, C. and Venter, J. C. (1992). Sequence identification of 2,375 human brain genes [see comments]. *Nature* **355**, 4632–4634.

Adams, M. D., Soares, M. B., Kerlavage, A. R., Fields, C. and Venter, J. C. (1993). Rapid CDNA sequencing (expressed sequence tags) from a directionally cloned human infant brain CDNA library. *Nat. Genet.* **4**, 373–380.

Adams, M. D. *et al.* (1994). *Automated DNA Sequencing and Analysis* (M. D. Adams, C. Fields, J. C. Venter, eds.). Academic Press, London, San Diego.

Ahringer, J. (1997). Turn to the Worm! *Curr. Opin. Genet. Dev.* **7**, 410–415.

Arlinghaus, H. F., Kwoka, M. N., Guo, X. Q. and Jacobson, K. B. (1997). Multiplexed DNA sequencing and diagnostics by hybridization with enriched stable isotope labels. *Anal. Chem.* **69**, 510–517.

Benbow, R. M. (1992). Chromosome structures. *Sci. Prog.* **76**, 425–450.

Benson, D. A., Boguski, M. S., Lipman, D. J., Ostell, J. and Ouellette, B. F. F. (1998). GenBank. *Nucl. Acids Res.* **26**, 1–7.

Bevan, M., Bancroft, I., Bent, E., Love, K., Goodman, H., Dean, C., Bergkamp, R., Dirkse, W., Van Staveren, M., Stiekema, W., Drost, L., Ridley, P., Hudson, S. A., Patel, K., Murphy, G., Piffanelli, P., Wedler, H., Wedler, E., Wambutt, R., Weitzenegger, T., Pohl, T. M., Terryn, N., Gielen, J., Villarroel, R., Chalwatzis, N. *et al.* (1998). Analysis of 1.9 Mb of contiguous sequence from chromosome 4 of *Arabidopsis thaliana* [in process citation]. *Nature* **391**, 485–488.

Bonfield, J. K., Smith, K. F. and Staden, R. (1995). A new DNA sequence assembly program. *Nucl. Acids Res.* **23**, 4992–4999.

Butler, B. (1994). Nucleic acid sequence analysis software packages. *Curr. Opin. Biotechnol.* **5**, 19–23.

Charmley, P., Nickerson, D. and Hood, L. (1994). Polymorphism detection and sequence analysis of human T-cell receptor V alpha-chain-encoding gene segments. *Immunogenetics* **39**, 138–145.

Cherry, J. M., Adler, C., Ball, C., Chervitz, S. A., Dwight, S. S., Hester, E. T., Jia, Y., Juvik, G., Roe, T., Schroeder, M., Weng, S. and Botstein, D. (1998). SGD: *Saccharomyces* genome database. *Nucl. Acids Res.* **26**, 73–79.

Civitello, A. B., Richards, S. and Gibbs, R. A. (1992). A simple protocol for the automation of DNA cycle sequencing reactions and polymerase chain reactions. *DNA Seq.* **3**, 17–23.

Dear, S. and Staden, R. (1991). A sequence assembly and editing program for efficient management of large projects. *Nucl. Acids Res.* **19**, 3907–3911.

Deininger, P. L. (1983). Random subcloning of sonicated DNA: application to shotgun DNA sequence analysis. *Anal. Biochem.* **129**, 216–223.

Deciphering Genomes

Drmanac, R., Drmanac, S., Labat, I., Crkvenjakov, R., Vicentic, A. and Gemmell, A. (1992). Sequencing by hybridization: towards an automated sequencing of one million M13 clones arrayed on membranes. *Electrophoresis* **13**, 566–573.

Drmanac, R., Drmanac, S., Strezoska, Z., Paunesku, T., Labat, I., Zeremski, M., Snoddy, J., Funkhouser, W. K., Koop, B., Hood, L. *et al.* (1993). DNA sequence determination by hybridization: a strategy for efficient large-scale sequencing [Published erratum appeared in *Science* 1994, **163**, 596]. *Science* **260**, 1649–1652

Du, Z., Hood, L. and Wilson, R. K. (1993). Automated fluorescent DNA sequencing of polymerase chain reaction products. *Meth. Enzymol* **218**, 104–121.

Edwards, A., Voss, H., Rice, P., Civitello, A., Stegemann, J., Schwager, C., Zimmermann, J., Erfle, H., Caskey, C. T. and Ansorge, W. (1990). Automated DNA sequencing of the human HPRT locus. *Genomics* **6**, 593–608.

Epplen, C., Santos, E. J., Maueler, W., van Helden, P. and Epplen, J. T. (1997). On simple repetitive DNA sequences and complex diseases. *Electrophoresis* **18**, 1577–1585.

Fullerton, S. M., Harding, R. M., Boyce, A. J. and Clegg, J. B. (1994). Molecular and population genetic analysis of allelic sequence diversity at the human beta-globin locus. *Proc. Natl. Acad. Sci. USA* **91**, 1805–1809.

Gerhold, D. and Caskey, C. T. (1996). It's the genes! EST access to human genome content. *Bioessays* **18**, 973–981.

Glover, R. P., Sweetman, G. M., Farmer, P. B. and Roberts, G. C. (1995). Sequencing of oligonucleotides using high performance liquid chromatography and electrospray mass spectrometry. *Rapid Commun. Mass Spectrom.* **9**, 897–901.

Goffeau, A., Barrell, B. G., Bussey, H., Davis, R. W., Dujon, B., Feldmann, H., Galibert, F., Hoheisel, J. D., Jacq, C., Johnston, M., Louis, E. J., Mewes, H. W., Murakami, Y., Philippsen, P., Tettelin, H. and Oliver, S. G. (1996). Life with 6,000 genes. *Science* **274**, 563–567.

Goldberg, R. B. (1978). DNA sequence organization in the soybean plant. *Biochem. Genet.* **16**, 45–68.

Goodman, H. M., Ecker, J. R. and Dean, C. (1995). The genome of *Arabidopsis thaliana*. *Proc. Natl. Acad. Sci. USA* **92**, 10831–10835.

Gurley, W. B., Hepburn, A. G. and Key, J. L. (1979). Sequence organization of the soybean genome. *Biochim. Biophys. Acta* **561**, 167–183.

Heiner, C. and Hunkapiller, T. (1989) Automated DNA sequencing. In *Nucleic Acids Sequencing: A Practical Approach* (C. J. Howe and E. S. Ward, eds), pp. 234–235. IRL Press, Oxford, England.

Hengen, P. N. (1997). Shearing DNA for genomic library construction. *Trends Biochem. Sci.* **22**, 273–274.

Holmquist, G. P. (1989). Evolution of chromosome bands: molecular ecology of noncoding DNA. *J. Mol. Evol.* **28**, 469–486.

Hood, L. E., Hunkapiller, M. W. and Smith, L. M. (1987). Automated DNA sequencing and analysis of the human genome. *Genomics* **1**, 201–212.

Hsu, T. C., Spirito, S. E. and Pardue, M. L. (1975). Distribution of 18+28S ribosomal genes in mammalian genomes. *Chromosoma* **53**, 25–36.

Huang, G. M., Wang, K., Kuo, C., Paeper, B. and Hood, L. (1994). A high-throughput plasmid DNA preparation method. *Anal. Biochem.* **223**, 35–38.

Hudson, T. J., Stein, L. D., Gerety, S. S., Castle, A. B., Silva, J., Slonim, D. K., Baptista, R., Kruglyak, L., Xu, S. H. *et al.* (1995). An STS-based map of the human genome. *Science* **270**, 1945–1954.

Hung, S. C., Ju, J., Mathies, R. A. and Glazer, A. N. (1996). Energy transfer primers with 5- or 6-carboxyrhodamine-6G as acceptor chromophores. *Anal. Biochem.* **238**, 165–170.

Hunkapiller, T. and Hood, L. (1991). LIMS and the human genome project. *Biotechnology* **9**, 1344–1345.

Ju, J., Glazer, A. N. and Mathies, R. A. (1996a). Energy transfer primers: a new fluorescence labeling paradigm for DNA sequencing and analysis. *Nat. Med.* **2**, 4347–4351.

Ju, J., Glazer, A. N. and Mathies, R. A. (1996b). Cassette labeling for facile construction of energy transfer fluorescent primers. *Nucl. Acids Res.* **24**, 246–249.

Jurka, J., Walichiewicz, J. and Milosavljevic, A. (1992). Prototypic sequences for human repetitive DNA. *J. Mol. Evol.* **35**, 286–291.

Kawasaki, K., Minoshima, S., Nakato, E., Shibuya, K., Shintani, A., Schmeitz, J. L., Wang, J. and Shimizu, N. (1997). One-megabase sequence analysis of the human immunoglobulin λ gene locus. *Genome Res.* **7**, 258–261.

Klenow, H. and Henningsen, I. (1970). Selective elimination of the exonuclease activity of the deoxyribonucleic acid polymerase from *Escherichia coli* B by limited proteolysis. *Proc. Natl. Acad. Sci. USA* **65**, 168–175.

Koop, B. F., Rowan, L., Chen, W. Q., Deshpande, P., Lee, H. and Hood, L. (1993). Sequence length and error analysis of sequenase and automated taq cycle sequencing methods. *BioTechniques* **14**, 442–447.

Kuwabara, P. E. (1997). Worming your way through the genome. *Trends Genet.* **13**, 455–460.

Lawrence, C. B., Honda, S., Parrott, N. W., Flood, T. C., Gu, L., Zhang, L., Jain, M., Larson, S. and Myers, E. W. (1994). The genome reconstruction manager: a software environment for supporting high-throughput DNA sequencing. *Genomics* **23**, 192–201.

Lee, L. G., Spurgeon, S. L., Heiner, C. R., Benson, S. C., Rosenblum, B. B., Menchen, S. M., Graham, R. J., Constantinescu, A., Upadhya, K. G. and Cassel, J. M. (1997). New energy transfer dyes for DNA sequencing. *Nucl. Acids Res.* **25**, 2816–2822.

Mardis, E. R. and Roe, B. A. (1989). Automated methods for single-stranded DNA isolation and dideoxynucleotide DNA sequencing reactions on a robotic workstation. *BioTechniques* **7**, 840–850.

McCormick, M. K., Buckler, A., Bruno, W., Campbell, E., Shera, K., Torney, D., Deaven, L. and Moyzis, R. (1993). Construction and characterization of a YAC library with a low frequency of chimeric clones from flow-sorted human chromosome 9. *Genomics* **18**, 553–558.

Mefford, H., van den Engh, G., Friedman, C. and Trask, B. J. (1997). Analysis of the variation in chromosome size among diverse human populations by bivariate flow karyotyping. *Hum. Genet.* **100**(1), 138–144.

Miller, M. J. and Powell, J. I. (1994). A quantitative comparison of DNA sequence assembly programs. *J. Comput. Biol.* **1**, 257–269.

Nickerson, D. A., Whitehurst, C., Boysen, C., Charmley, P., Kaiser, R. and Hood, L. (1992). Identification of clusters of biallelic polymorphic sequence-tagged sites (PSTSs) that generate highly informative and automatable markers for genetic linkage mapping. *Genomics* **12**, 377–387.

Olson, M., Hood, L., Cantor, C. and Botstein, D. (1989). A common language for physical mapping of the human genome. *Science* **245**, 1434–1435.

Parker, S. R. (1997). Autoassembler sequence assembly software. *Meth. Mol. Biol.* **70**, 107–117.

Parsons, J. D. (1995). Miropeats: graphical DNA sequence comparisons. *Comput. Appl. Biosci.* **11**, 615–619.

Reeve, M. A. and Fuller, C. W. (1995). A novel thermostable polymerase for DNA sequencing. *Nature* **376**, 796–797.

Report of the Task Force on Genetic Information and Insurance (1993). Genetic Information and Health Insurance. NIH/DOE Working Group on Ethical, Legal, and Social Implications of Human Genome Research. *Hum. Gene Ther.* **4**, 789–808.

Rieder, M. J., Taylor, S. L., Tobe, V. O. and Nickerson, D. A. (1998). Automating the identification of DNA variations using quality-based fluorescence re-sequencing: analysis of the human mitochondrial genome. *Nucl. Acids Res.* **26**, 967–973.

Roach, J. C. (1995). Random subcloning. *Genome Res.* **5**, 464–473.

Roach, J. C., Boysen, C., Wang, K. and Hood, L. (1995). Pairwise end sequencing: a unified approach to genomic mapping and sequencing. *Genomics* **26**, 345–353.

Rosenblum, B. B., Lee, L. G., Spurgeon, S. L., Khan, S. H., Menchen, S. M., Heiner, C. R. and Chen, S. M. (1997). New dye-labeled terminators for improved DNA sequencing patterns. *Nucl. Acids Res.* **25**, 4500–4504.

Rowen, L. and Koop, B. (1994). *Automated DNA Sequencing and Analysis* (M. D. Adams, C. Fields and J. C. Venter, eds), pp. 167–174. Academic Press, London, San Diego.

Rowen, L., Koop, B. F. and Hood, L. (1996). The complete 685-kilobase DNA sequence of the human beta T cell receptor locus. *Science* **272**, 1755–1762.

Rowen, L., Mahairas, G. and Hood, L. (1997). Sequencing the human genome. *Science* **278**, 605–607.

Sanger, F., Nicklen, S. and Coulson, A. R. (1977). DNA sequencing with chain-terminating inhibitors. *Proc. Natl. Acad. Sci. USA* **74**, 5463–5467.

Sargent, R., Fuhrman, D., Critchlow, T., Di Sera, T., Mecklenburg, R. and Cartwright, P. (1996). The design and implementation of a database for human genome research. Eighth International Conference on Scientific and Statistical Database Management, IEEE Computer Society Press, Stockholm, Sweden.

Selleri, L., Eubanks, J. H., Giovannini, M., Hermanson, G. G., Romo, A., Djabali, M., Maurer, S., McElligott, D. L., Smith, M. W. and Evans, G. A. (1992). Detection and characterization of "chimeric" yeast artificial chromosome clones by fluorescent in situ suppression hybridization. *Genomics* **14**, 536–541.

Shizuya, H., Birren, B., Kim, U. J., Mancino, V., Slepak, T., Tachiiri, Y. and Simon, M. (1992). Cloning and stable maintenance of 300-kilobase-pair fragments of human DNA in *Escherichia coli* using an F-factor-based vector. *Proc. Natl. Acad. Sci. USA* **89**, 8794–8797.

Slightom, J. L., Siemieniak, D. R., Sieu, L. C., Koop, B. F. and Hood, L. (1994). Nucleotide sequence analysis of 77.7 Kb of the human V beta T-cell receptor gene locus: direct primer-walking using cosmid template DNAs. *Genomics* **20**, 149–168.

Smit, A. F. (1996). The origin of interspersed repeats in the human genome. *Curr. Opin. Genet. Dev.* **6**, 743–748.

Smith, L. M. (1989). Automated DNA sequencing and the analysis of the human genome. *Genome* **31**, 929–937.

Smith, L. M., Sardens, J. Z., Kaiser, R. J., Hughes, P., Dodd, C., Connell, C., Hienre, C., Kent, S. B. H. and Hood, L. E. (1986). Fluorescense detection in automated sequence analysis. *Nature* **321**, 674–679.

Smith, T. M., Abajian, C. and Hood, L. (1997). Hopper: software for automating data tracking and flow in DNA sequencing. *Comput. Appl. Biosci.* **13**, 175–182.

Strezoska, Z., Paunesku, T., Radosavljevic, D., Labat, I., Drmanac, R. and Crkvenjakov, R. (1991). DNA sequencing by hybridization: 100 bases read by a non-gel-based method. *Proc. Natl. Acad. Sci. USA* **88**, 10089–10093.

Swindell, S. R. and Plasterer, T. N. (1997). SEQMAN. Contig assembly. *Meth. Mol. Biol.* **70**, 75–89.

Tabor, S. and Richardson, C. C. (1987). DNA sequence analysis with a modified bacteriophage T7 DNA polymerase. *Proc. Natl. Acad. Sci. USA* **84**, 4767–4771.

Tabor, S. and Richardson, C. C. (1995). A single residue in DNA polymerases of the *Escherichia coli* DNA polymerase I family is critical for distinguising between deoxy- and dideoxyribonucleotides. *PNAS* **92**(14), 6339–6343.

Trask, B. J., Friedman, C., Martin-Gallardo, A., Rowen, L., Akinbami, C., Blankenship, J., Collins, C., Giorgi, D., Iadonato, S., Johnson, F., Kuo, W. L., Massa, H., Morrish, T., Naylor, S., Nguyen, O. T. H., Rouquier, S., Smith, T., Wong, D. J., Youngblom, J. and van den Engh, G. (1998). Members of the olfactory receptor gene family are contained in large blocks of DNA duplicated polymorphically near the ends of human chromosomes. *Hum. Mol. Genet.* **7**, 13–26.

Venter, J. C., Smith, H. O. and Hood, L. (1996). A new strategy for genome sequencing [see comments]. *Nature* **381**, 364–366.

Voss, H., Schwager, C., Wiemann, S., Zimmermann, J., Stegemann, J., Erfle, H., Voie, A. M., Drzonek, H. and Ansorge, W. (1995). Efficient low redundancy large-scale DNA sequencing at EMBL. *J. Biotechnol.* **41**, 121–129.

Wada, M., Abe, K., Okumura, K., Taguchi, H., Kohno, K., Imamoto, F., Schlessinger, D. and Kuwano, M. (1994). Chimeric YACs were generated at unreduced rates in conditions that suppress coligation. *Nucl. Acids Res.* **22**, 1651–1654.

Wilson, R. K. (1993). High-throughput purification of M13 templates for DNA sequencing. *BioTechniques* **15**, 414–416.

Wong, G. K., Yu, J., Thayer, E. C. and Olson, M. V. (1997). Multiple-complete-digest restriction fragment mapping: generating sequence-ready maps for large-scale DNA sequencing. *Proc. Natl. Acad. Sci. USA* **94**, 5225–5330.

◆◆◆◆◆◆ NOTE ADDED IN PROOF

NIH and DOE have made a recent proposal to cover this portion of the human genome (~ 60%) with a low-pass sequencing effort by the year 2001 (Marshall, E., 1998; NIH to produce a "Working Draft" of the Genome by 2001. *Science* **281**, 1774–1775).

8 DNA Arrays for Transcriptional Profiling

Nicole C. Hauser[1], Marcel Scheideler[1], Stefan Matysiak[1], Martin Vingron[2] and Jörg D. Hoheisel[1]

[1] *Functional Genome Analysis, Deutsches Krebsforschungszentrum, Heidelberg, Germany;* [2] *Theoretical Bioinformatics, Deutsches Krebsforschungszentrum, Heidelberg, Germany*

CONTENTS

I. INTRODUCTION

Recently, interest in a whole-genome analysis of micro-organisms and, as a first step thereof, the sequencing has taken a leap forward. Although initially also spurred by the mere congruence between genome size and sequencing capacity, the sequencing of microbial genomes is now recognised as an integral part of genome research, at present probably producing more data of biological and medical consequence per base pair than sequencing projects on higher organisms. Currently, the finished sequences of 16 microbial genomes are available in the public domain (www.tigr.org/tdb/mdb/mdb.html) and 80 or so are under way, with even more to come. Already, however, the next phases have started towards a real understanding of intracellular activity on a molecular level (Oliver, 1996), with microbial systems in this respect again acting as a testing ground for technologies that eventually will be used for analyses on higher organisms.

One essential aspect of such studies is the investigation of gene activity on three levels – promoter activity, RNA stability and subsequent translation into protein – and the regulation of these stages of expression. Transcriptional analysis by hybridising complex RNA probes to arrays made of gene representatives permits studies on two of the above-listed issues, although reduced to a single measure, determined as the actual amount of RNA present in cells at a given point. Sample hybridisation to gene arrays, which until recently was mostly carried out using anonymous

METHODS IN MICROBIOLOGY, VOLUME 28
0580-9517 $30.00

cDNA sequences, has indicated the usefulness of the information acquired by this means (Augenlicht *et al.*, 1991; DeRisi *et al.*, 1996; Gress *et al.*, 1992; Höög, 1991; Nguyen *et al.*, 1995, Schena *et al.*, 1995). With the availability of complete, non-redundant gene repertoires, however, a new quality level has been reached based on which even complex analyses could be performed (DeRisi *et al.*, 1997; de Saizieu *et al.*, 1998; Hauser *et al.*, 1998; Lashkari *et al.*, 1997; Wodicka *et al.*, 1997). The whole benefit of such studies will only become really apparent, when such data is merged with other data resulting from very different analyses, such as biochemical assays, for example, to produce added value by comparison and parallel evaluation. Additionally, DNA chips as a common analytical tool will be an important factor in linking the different, specific fields of interest: data will be used in the particular investigation and assist in the particular analysis while, concurrently, data cross-referencing will interrelate individual aspects to give a better understanding of the whole picture.

◆◆◆◆◆◆ II. RESULTS

A. Spot Density and Support Media

Within the European Functional Analysis Network (EUROFAN), we have embarked on the distribution of arrays that consist of the complete set of yeast genes. These arrays only represent a mark two generation, however. While containing basically all detected open reading frames (ORFs) as PCR products rather than *in situ* blotted *E. coli* cells containing cDNA clones – the initial format used in large-scale analyses (Lehrach *et al.*, 1990; Poustka *et al.*, 1986) – the spot density is relatively low, restrained by the lack of high-resolution detection units in the receiving laboratories. The arrays distributed have the PCR products placed at a density compatible with the relatively widely disseminated phosphor- and fluoroimager technology, equipment with pixel sizes ranging between 50 and 100 μm. For yeast, and many other microbial and especially bacterial organisms, this does not matter too much, since usually the amount of probe material is not limited. Therefore, even nylon filters with densities of around 60 DNA spots per centimetre square are sufficient for many analyses.

Using commercial robotic devices (e.g. BioRobotics, UK) equipped with a 384-pin tool, the approximately 6200 PCR products made of the yeast ORFs were arrayed on an area equivalent to three microtitre dishes, each fragment being present twice. Using the same basic set-up but applying a newly designed micro-pin tool (BioRobotics, UK), all fragments could be placed on an area equivalent to one microtitre dish. With a commercial device based on piezo pipetting technology (GeSiM, Germany) for the application of PCR fragments from 96-well or 384-well plates, respectively, the distances between spots could be reduced significantly and less material had to be transferred. Because of the high spatial resolution and the reproducibly small drop volume, chip densities of up to 10 000 spots per centimetre square – and more with further reduced drop sizes – are possible with this device; other laboratories working on similar approaches have achieved even higher densities using self-made spotting equipment for the fabrication of the arrays (Lashkari *et al.*, 1997; Yershov *et al.*, 1996) or taking advantage of *in situ* oligomer synthesis controlled by photolithographic techniques (Wodicka *et al.*, 1997). All real chip applications, however, are currently restricted in their dissemination and take place *only* at a few central facilities, since especially at academic institutions the appropriate reading devices are missing. This, however, will change soon with purpose-built machines becoming commercially available at reasonable cost.

When nylon filters are the support medium the sensitivity issue may be problematic. In accordance with results reported earlier (Nguyen *et al.*, 1995), we found that a signal originating from individual transcripts that each represent around 0.01% of a total mRNA mixture is about the best one can expect (Hauser *et al.*, 1998). The relatively high background typical of nylon filters is limiting. As determined by covalently binding labelled DNA to a filter prior to hybridisation, signals of low intensity were simply submerged by noise. Hence, increasing the probe concentration does not improve the results. Mainly for this reason, other, more inert surfaces are advantageous, such as glass (DeRisi *et al.*, 1997; Maskos and Southern, 1992) or polypropylene (Matson *et al.*, 1995; Weiler and Hoheisel, 1996), which exhibit only little unspecific binding of a probe. An increase in probe concentration therefore translates directly into higher sensitivity.

B. Array-bound Molecules

To date, the most common detector molecules attached to the solid support were either material from cDNA clone colonies, PCR products made of cDNA inserts or material amplified directly from the genomic DNA, the last obviously dependent on the availability of the genomic sequence. The presence of the large amount of *E. coli* DNA as part of the *in situ* attached DNA from clone colonies amounts to a considerable problem because of interactions, specific and unspecific, with the probe molecules. Especially when using high probe concentrations needed to detect rare transcripts, the background increases significantly. By placing PCR products on to the arrays, such problems are avoided. Since there is no system yet for high-throughput preparations of (plasmid) DNA, PCR also provides the means for the purification of very many samples (Figure 1). In addition, probe can be isolated both from cDNA clones and directly from genomic DNA.

Disadvantages of using an entire gene sequence for detection are the inability to analyse exon-specific effects and the relatively small selectivity of hybridisation, causing cross-hybridisation between homologous sequences. While the former is not much of a problem for microbial organisms with no or only very short introns being present, the latter will lead to erroneous data interpretation in so far as relatively short but highly conserved DNA domains will give rise to signals at DNA samples that might otherwise be entirely unrelated sequences.

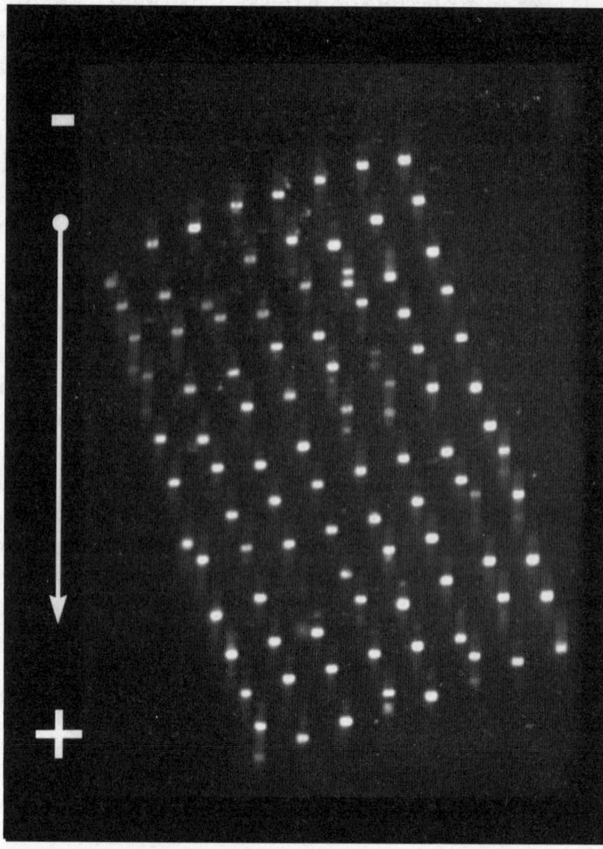

Figure 1. Gel electrophoretic separation of 96 typical PCR products out of a total of 13 300 independent amplifications of a gene set of a single organism, each done in a volume of 100 µl of which 5 µl was loaded to the gel. Only few of the amplifications were unsuccessful, indicated by the lack of any product and the concomitant presence of large primer quantities.

Support-attached oligonucleotides (de Saizieu *et al.*, 1998; Lockhart *et al.*, 1996; Wodicka *et al.*, 1997; Yershov *et al.*, 1996) permit highly discriminative hybridisation. The high degree of binding selectivity, however, is achieved at a cost in stability of the duplexes formed between the oligonucleotides and the probe molecules. Thus, high probe concentrations are required in order to achieve good signal intensities. Also, the probe has to be fragmented to avoid the formation of secondary structures that would hamper binding to the arrayed oligonucleotides (Milner *et al.*, 1997). Overall, the usage of oligonucleotide arrays is currently much more costly than the use of PCR products, if only because large numbers of oligonucleotides have to be synthesised (on the chip or elsewhere), and sequence has to be known in the first place. Such is also necessary for PCR amplification from genomic DNA, but only two primers would be required rather than the numerous oligonucleotides applied to the chips for reasons of quality control.

Technically improved analysis based on oligomer sequences should be possible by using peptide nucleic acid (PNA) oligomers as substrate on the arrays (Weiler *et al.*, 1997, 1998), merging the advantages of the above approaches. For yeast, we are currently working on a comprehensive set of PNA oligomers. Several features of PNA foster superior results. Duplex stability of PNA:DNA or PNA:RNA hybrids is high, with dissociation temperatures of 16-mer sequences being in a range of 60 to 80°C. Nevertheless, PNA oligomers, in most cases, exhibit an even higher selectivity than DNA oligonucleotides, let alone PCR products. Probe accessibility is better, since intramolecular folding of the probe is diminished because of a very low ion concentration needed in the hybridisation buffer. Since PNA is an uncharged molecule, no ions are required for counteracting inter-strand repulsion between annealing molecules. Finally, PNA can invade double-stranded nucleic acids by replacing one strand while binding to its complementary sequence.

C. Probe Generation

The method used for RNA isolation was found to be critical for the success of our analyses. Several preparations obtained from different sources yielded only insufficient probe, although the quality seemed to be good as judged by OD measurement and gel analysis. Also, in our personal experience, isolation of RNA by phenol and chloroform extractions, for example, only produced RNA of variable quality for reverse transcription. No obvious reason for the high degree of variability could be identified. Another problem encountered was the issue of probe characteristics. A standard protocol for yeast relies on the generation of protoplasts prior to the actual RNA extraction, for instance. Such treatment provided RNA that worked very well for probe generation, but the technique induced an intracellular stress reaction during the process of cell wall removal, whereby transcriptional activities were strongly influenced. While this effect could be circumvented by simply freezing the yeast cells immediately after growth, it highlights the importance of taking into account the cell harvesting and RNA isolation procedures in order to avoid or at least minimise the risk of artificially causing transcriptional responses which have more to do with the experimental manipulations rather than the culture conditions.

Our eventual procedure for RNA extraction relied on the use of a monophasic solution of phenol and guanidine isothiocyanate and proved robust in both aspects, i.e. unbiased RNA levels and good probe generation. Cells were instantly shock-frozen by directly releasing drops of the growth culture into liquid nitrogen in the small Teflon vessel of a microdismembrator (Braun Melsungen, Germany), kept frozen during mechanical breakage and only thawed when suspended in the organic solvent. By this method, some 250 µg of RNA were obtained within 2 hours from 15 OD_{600} units of yeast cells, for example (Hauser *et al.*, 1998).

For labelling, RNA was reverse transcribed in the presence of a large excess of oligo-dT primer molecules as described (Nguyen *et al.*, 1995). This procedure is optimised for minimising the portion of poly-A sequence that is reverse transcribed. With no poly-(A:T) sequences present on the yeast arrays, this fact was inconsequential with respect to the specificity of hybridisation but meaningful nevertheless for a reduction of any bias introduced by potential priming differences caused by a transcript's tail length. Variations in the effectiveness of the labelling procedure were assessed by adding, as a control, a known amount of mRNA of the rabbit β-globin gene (Life Technologies, UK) to RNA isolates prior to the probe preparation. With the rabbit gene present on an array, trans-effects could be checked for directly.

In organisms without polyadenylated mRNA, total RNA is to be used as probe, either reverse transcribed by random priming or directly labelled (de Saizieu *et al.*, 1998). Although the large percentage of ribosomal RNA present in such a probe is bound to increase background while simultaneously diluting the specific activity of the mRNA-complementary portion, there should be sufficient material nevertheless. Alternatively, the ribosomal RNA contamination could be reduced by subtraction protocols (e.g. Geng *et al.*, 1998; Korn *et al.*, 1992). It still has to be demonstrated, however, that these procedures do not introduce a bias by means of the manipulation procedures involved.

D. Detection

Currently, there are two labelling methods routinely used for probe detection, although others are possible and being researched. One is the use of radioactivity, in particular ^{33}P due to its superiority with regard to the resolution of detection compared to the cheaper ^{32}P label. For all its drawbacks, radioactive labelling represents a system that is sensitive, well proven and established, important factors when it comes to quantitative analyses. The replacement of radioactivity by a fluorescently labelled probe is a prerequisite for analysis on a chip. The parallel use of two different dyes permits immediate internal controls, and optical systems allow for very high spatial precision. A requirement for this, however, is that detection systems of sufficient sensitivity and resolution are actually available (Figure 2). The technology of optical detection has further to go. For improved results, excitation could be done by taking advantage of optical wave guides, for example, which cause excitation to take place only in the zone reached by the evanescent wave along the wave guide material (Stimpson *et al.*, 1995). Thus, only label that is very close to the support, nearly exclusively bound probe molecules, will produce a signal.

E. Experimental Reproducibility

Our arrays both on nylon and polypropylene were used several times with the same probe to check for the reproducibility of the transcriptional results. Comparing hybridisations performed with a 5′-tag sequence common to all PCR primers, carried out before and after a set of complex hybridisations, the typical correlation coefficient was found to be 0.99, indicating that the amount of DNA (and thus signal intensity) at each spot

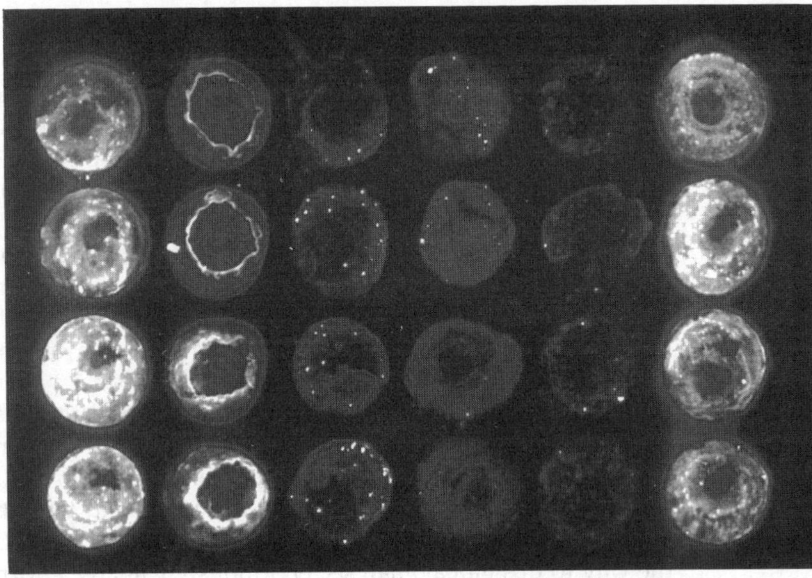

Figure 2. Detection of spots containing a dilution of fluorescence material. Each spot has a diameter of about 250 µm. In the weakest column (second from the right), each spot represents 10^{-19} mol of material (done in collaboration with Josef Atzler, Markus Rauchensteiner und Daniel Stock of TECAN Austria).

remained constant over this experimental setting (Hauser *et al.*, 1998). Moreover, these data were used for normalising the DNA amounts present at each spot. When data sets from actual transcriptional analyses with RNA from identical samples were measured in different experiments, typically a correlation coefficient of 0.97 was obtained demonstrating the high degree of reproducibility of even complex data in duplicate experiments. Nevertheless, the experimental variation was such that the average of at least two identical assays was taken into account in actual analyses. In this respect, the reusability of the arrays is an important issue and a critical aspect in our philosophy of chip production.

F. Data analysis

Analysis of the huge amount of raw data generated by the type of experiment described here is still in its infancy. For our studies, a software package was written (Hauser *et al.*, 1998; Vingron *et al.*, in prep.), and work is continuing. Most features implemented so far deal more with data assessment and presentation rather than addressing the issue of filtering out the information that could be relevant to the addressed problem. Only slowly, the correct sort of algorithms beyond the obvious start to emerge and take shape. However, a sensible merging of the transcriptional profiling data with the results from other areas of analysis will eventually provide the means to ask the right sort of questions and retrieve the appropriate answers.

◆◆◆◆◆◆ III. DISCUSSION

A simultaneous analysis of the expression level on *all* genes of an organism is a prerequisite to the understanding of regulative cellular processes. Currently, for such assays, DNA chips present the best methodology of accumulating the amount of transcriptional information necessary to unravel the complex connections. The basic technology is at hand and will develop further. RNA isolates from cultures or tissues grown at various conditions or treated with certain compounds will provide information on gene activity and regulation; in more medical terms, the activity of drugs or adaptations of microbial systems to such treatment could be analysed on a molecular level, even for individual patients. For high-throughput screening, a mere pattern comparison rather than precise signal quantification at the individual spots could be very rewarding. Ultimately, the technology will be used as a system to test many biological effects on a molecular basis and to understand the interactions that take place within the complex regulation circuits of a cell. But already on the way to this global analysis, enormously valuable information will be extracted. One should keep in mind, however, that transcriptional data are only a relatively small part of the whole picture. Regulation patterns will only be really understood, even in simple terms, if promoter activity, transcriptional data and actual protein expression levels are readily available for comparison. But even then, effects caused from protein modification and interaction or cell compartmentation cannot be assayed as such.

In current set-ups, the whole genome sequencing still is preceding large-scale functional analyses. This is bound to change, however. With the large increase in sequence capacity and willingness, not to say eagerness, to analyse microbial organisms in much detail, it will not be long before a sequence analysis of yet another micro-organism will mostly produce redundant information that could have been deduced from already existing data by an *in silico* analysis. Then, a large portion of a sequencing effort would be squandered, which even in times of improved sequence technology is still a considerable waste of time and money. A combination of existing techniques could be the way out of this dilemma, actually turning on its head the order of analysis, making functional analyses the foundation of any subsequent sequencing (Figure 3).

Generation of high-resolution physical maps made from shotgun template clones for low-redundant genomic sequencing was a technical development of the yeast sequencing programme (Johnston *et al.*, 1997; Scholler *et al.*, 1995). Even on more complex human DNA, template mapping has proved highly efficient at a current cost of around one-tenth that of sequencing and the cost is bound to fall with increased automation (Scholler *et al.*, 1998). Such analyses result in high-resolution maps which reflect genomic DNA at a resolution of approximately 200 bp. From such representations, a tiling path is selected for low-redundancy sequencing, with the additional advantage of a much simplified sequence assembly and finishing phase. For a microbial genome, such a set of genomic template clones also represents a condensed, normalised gene inventory and transcript map. Due to the high gene density, nearly every one of the 1 or

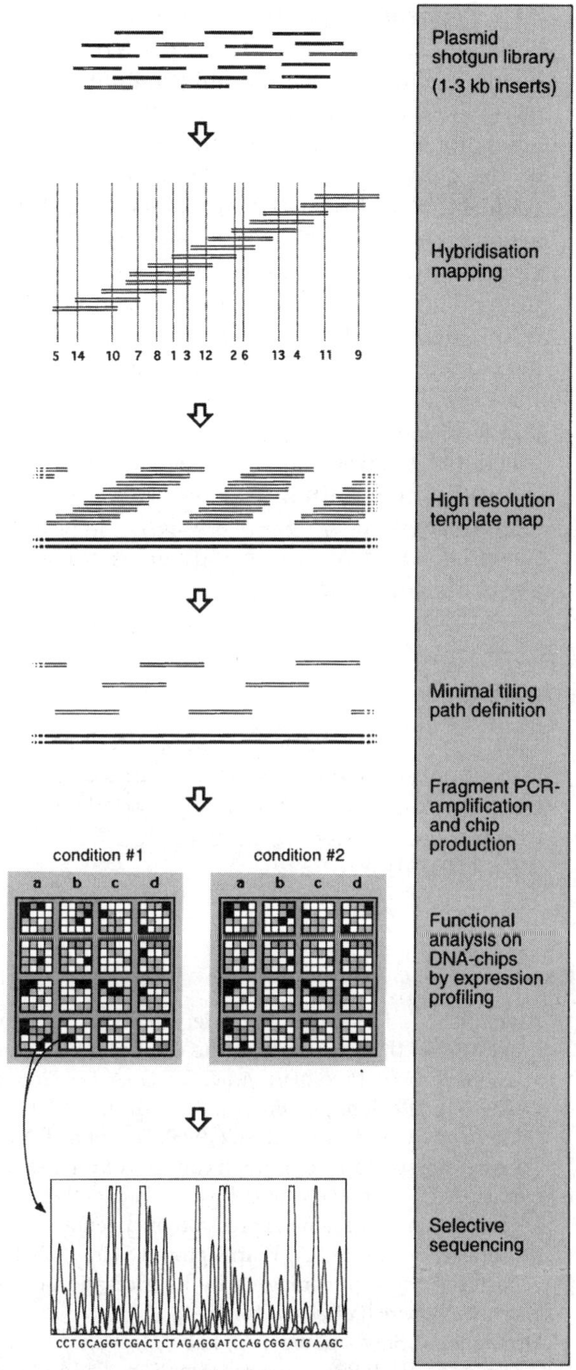

Figure 3. Scheme depicting the strategy of performing whole-genome functional analyses on microbial organisms prior to a selective sequencing of interesting regions. Details are given in the text.

3 kb fragments will contain at least part of a gene. Instead of proceeding with the sequence analysis of the ordered but still anonymous DNA fragments – although tag-like information could have been attributed by using motif oligonucleotides during mapping (Drmanac *et al.*, 1996, 1998) – functional assays on clone arrays of a minimally overlapping set should be the next step (Figure 3). PCR products of the individual fragments could be placed on DNA arrays as a substrate for the identification of genomic regions that exhibit interesting transcriptional response to a given stimulus. Also, comparative studies by hybridising genomic DNA from related organisms or mutant strains could be done, for example. On colony filters (Lehrach *et al.*, 1990), promoter activities could be tested, if the cloning vector would contain a suitable reporter gene (Niedenthal *et al.*, 1996), or Western-blot-like analyses on *in situ* expressed proteins might add relevant data. Only after the performance of these or other functional analyses, will the genes and related genomic regions that showed an interesting response be picked out for sequencing. By this selective approach, only a few tens of thousands of base pairs of high potential will have to be sequenced rather than millions of bases of unqualified importance.

Acknowledgements

This work was financially supported as part of the German Yeast Functional Analysis Network, funded by the German Science and Research Ministry (BMBF), and grants obtained from the European Commission under contracts BIO4-CT95-0080, BIO4-CT97-2294 and BIO4-CT95-0147.

References

Augenlicht, L. H., Taylor, J., Anderson, L. and Lipkin, M. (1991). Patterns of gene expression that characterise the colonic mucosa in patients at genetic risk for colonic cancer. *Proc. Natl. Acad. Sci. USA.* **88**, 3286–3289.

DeRisi, J. L., Penland, L., Brown, P. O., Bittner, M. L., Meltzer, P. S., Ray, M., Chan, Y., Su, Y. A. and Trent, J. M. (1996). Use of a cDNA microarray to analyse gene expression patterns in human cancer. *Nature Genet.* **14**, 457–460.

DeRisi, J. L., Iyer, V. R. and Brown, P. O. (1997). Exploring the metabolic and genetic control of gene expression on a genomic scale. *Science* **278**, 680–686.

de Saizieu, A., Certa, U., Warrington, J., Gray, C., Keck, W. and Mous, J. (1998). Bacterial transcript imaging by hybridisation of total RNA to oligonucleotide arrays. *Nature Biotechnol.* **16**, 45–48.

Drmanac, S., Stavropoulos, A., Labat, I., Vonau, J., Hauser, B., Soares, M. B. and Drmanac, R. (1996). Gene-representing cDNA clusters defined by hybridisation of 57,419 clones from infant brain libraries with short oligonucleotide probes. *Genomics* **37**, 29–40.

Drmanac, S., Kita, D., Labat, I., Hauser, B., Schmidt, C., Burczak, J.D. and Drmanac, R. (1998). Accurate sequencing by hybridisation for DNA diagnostics and individual genomics. *Nature Biotechnol.* **16**, 54–58.

Geng, M., Wallrapp, C., Müller-Pillasch, F., Frohme, M., Hoheisel, J. D. and Gress,

T. (1998). Isolation of differentially expressed genes by combining representational difference analysis (RDA) and cDNA library arrays. *Biotechnique* (in press).

Gress, T. M., Hoheisel, J. D., Lennon, G. G., Zehetner, G. and Lehrach, H. (1992). Hybridisation fingerprinting of high density cDNA-library arrays with cDNA pools derived from whole tissues. *Mamm. Genome* **3**, 609–619.

Hauser, N. C., Vingron, M., Scheideler, M., Krems, B., Hellmuth, K., Entian, K.-D. and Hoheisel, J. D. (1998). Transcriptional profiling on all open reading frames of *Saccharomyces cerevisiae*. *Yeast* **14,** 1209–1221.

Höög, C. (1991). Isolation of large number of novel mammalian genes by a differential cDNA library screening strategy. *Nucl. Acids Res.* **19**, 6123–6127.

Johnston, M., Hillier, L., Riles, L. *et al.* (1997). The nucleotide sequence of *Saccharomyces cerevisiae* chromosome XII. *Nature* **387** (suppl.), 87–90.

Korn, B., Sedlacek, Z., Manca, A., Kioschis, P., Konecki, D., Lehrach, H. and Poustka, A. (1992). A strategy for the selection of transcribed sequences in the Xq28 region. *Hum. Mol. Genet.* **1**, 235–242.

Lashkari, D. A., DeRisi, J. L., McCusker, J. H., Namath, A. F., Gentile, C., Hwang, S. Y., Brown, P. O. and Davis, R. W. (1997). Yeast microarrays for genome wide parallel genetic and gene expression analysis. *Proc. Natl. Acad. Sci. USA* **94**, 13057–13062.

Lehrach, H., Drmanac, R., Hoheisel, J. D., Larin, Z., Lennon, G., Monaco, A. P., Nizetic, D., Zehetner, G. and Poustka, A. (1990). Hybridisation fingerprinting in genome mapping and sequencing. In *Genome Analysis: Genetic and Physical Mapping* (K. E. Davies and S. Tilghman, eds), pp. 39–81. Cold Spring Harbor Laboratory Press, Cold Spring Harbor, New York.

Lockhart, D. L., Dong, H., Byrne, M. C., Follettie, M. T., Gallo, M. V., Chee, M. S., Mittmann, M., Wang, C., Kobayashi, M., Horton, H. and Brown, E. L. (1996). Expression monitoring by hybridisation to high-density oligonucleotide arrays. *Nature Biotechnol.* **14**, 1675–1680.

Maskos, U. and Southern, E. M. (1992). Oligonucleotide hybridisations on glass supports: a novel linker for oligonucleotide synthesis and hybridisation properties of oligonucleotides synthesised in situ. *Nucl. Acids Res.* **20**, 1679–1684.

Matson, R. S., Rampal, J., Pentoney, S. L., Anderson, P. D. and Coassin, P. (1995). Biopolymer synthesis on polypropylene support: oligonucleotide arrays. *Anal. Biochem.* **224**, 110–116.

Milner, N., Mir, K. U. and Southern, E. M. (1997). Selecting effective antisense reagents on combinatorial oligonucleotide arrays. *Nature Biotechnol.* **15**, 537–541.

Nguyen, C., Rocha, D., Granjeaud, S., Baldit, M., Bernard, K., Naquet, P. and Jordan, B. R. (1995). Differential gene expression in the murine thymus assayed by quantitative hybridisation of arrayed cDNA clones. *Genomics* **29**, 207–216.

Niedenthal, R. K., Riles, L., Johnston, M. and Hegemann, J. H. (1996). Green fluorescent protein as a marker for gene expression and subcellular localisation in budding yeast. *Yeast* **12**, 773–786.

Oliver, S. G., van der Aart Q. J., Agostoni-Carbone, M. L. *et al.* (1996). From DNA sequence to biological function. *Nature* **379**, 597–600.

Poustka, A., Pohl, T., Barlow, D. P., Zehetner, G., Craig, A., Michiels, F., Ehrich, E., Frischauf, A. M. and Lehrach, H. (1986). Molecular approaches to mammalian genetics. *Cold Spring Harb. Symp. Quant. Biol.* **51**, 131–139.

Schena, M., Shalon, D., Davis, R. W. and Brown, P. O. (1995). Quantitative monitoring of gene expression patterns with a complementary DNA microarray. *Science* **270**, 467–470.

Scholler, P., Karger, A. E., Meier-Ewert, S., Lehrach, H., Delius, H. and Hoheisel, J. D. (1995). Fine-mapping of shotgun template-libraries; an efficient strategy for the systematic sequencing of genomic DNA. *Nucl. Acids Res.* **23**, 3842–3849.

Scholler, P., Heber, S. and Hoheisel, J. D. (1998). Optimisation and automation of fluorescence-based DNA hybridisation for high-throughput clone mapping. *Electrophoresis* **19**, 504–508.

Stimpson, D. I., Hoijer, J. V., Hsieh, W.-T., Jou, C., Gordon, J., Theriault, T., Gamble, R. and Baldeschwieler, J. D. (1995). Real-time detection of DNA hybridisation and melting on oligonucleotide arrays by using optical wave guides. *Proc. Natl. Acad. Sci. USA* **92**, 6379–6383.

Weiler, J. and Hoheisel, J. D. (1996). Combining the preparation of oligonucleotide arrays and synthesis of high quality primers. *Anal. Biochem.* **243**, 218–227.

Weiler, J., Gausepohl, H., Hauser, N., Jensen, O. N. and Hoheisel, J. D. (1997). Hybridisation based DNA screening on peptide nucleic acid (PNA) oligonucleotide arrays. *Nucl. Acids Res.* **25**, 2792–2799.

Weiler, J., Matysiak, S., Gausepohl, H. and Hoheisel, J. D. (1998). New developments in hybridisation based DNA screening on oligomer arrays. In *Solid Phase Synthesis – Peptides, Proteins and Nucleic Acids* (R. Epton, ed.), Mayflower Worldwide, Oxford (in press).

Wodicka, L., Dong, H., Mittmann, M., Ho, M.-H. and Lockhart, D.J. (1997). Genome-wide expression monitoring in *Saccharomyces cerevisiae*. *Nature Biotechnol.* **15**, 1359–1367.

Yershov, G., Barsky, V., Belgovskiy, A., Kirillov, E., Kreindlin, E., Ivanov, I., Parinov, S., Guschin, D., Drobishev, A., Dubiley, S. and Mirzabekov, A. (1996). DNA analysis and diagnostics on oligonucleotide microchips. *Proc. Natl. Acad. Sci. USA* **93**, 4913–4918.

9 Large-scale Phenotypic Analysis in Microtitre Plates of Mutants with Deleted Open Reading Frames from Yeast Chromosome III: Key-step Between Genomic Sequencing and Protein Function

Klaus-Jörg Rieger[1], Gabriela Orlowska[2], Aneta Kaniak[1], Jean-Yves Coppee[1], Gordana Aljinovic[3] and Piotr P. Slonimski[1]

[1] Centre de Génétique Moléculaire du Centre National de la Recherche Scientifique, Laboratoire Propre Associé à l'Université Pierre et Marie Curie, Gif-sur-Yvette, France; [2] Institute of Microbiology, University of Wroclaw, Poland; [3] GATC-Gesellschaft für Analyse Technik und Consulting, Fritz-Arnold-Strasse, Konstanz, Germany

CONTENTS

I. INTRODUCTION

Five years ago, a consortium of 35 European laboratories established the first complete sequence of a eukaryotic chromosome, that of chromosome III from the budding yeast *Saccharomyces cerevisiae* (Oliver *et al.*, 1992). Recently, the yeast genome has been completely sequenced and the 6000 open reading frames (ORFs), potentially coding for proteins, have been identified (Goffeau *et al.*, 1996). The case for choosing yeast as the most appropriate organism to move into this new dimension of biological research is overwhelming (well-known eukaryote, compact genome, powerful classical and reverse genetics, numerous homologies to human genes, targeted gene disruption, large scientific community, possible industrial applications, no ethical concerns). Indeed, the systematic sequencing of the genome of this model organism opens the door to the identification of basic biological mechanisms common to all eukaryotes, including man, which are not accessible through classical approaches. The main finding of the sequencing project concerns the abundance of novel genes and gene families which was unexpected from the previous genetic and biochemical approaches. Indeed, about 50% of the new genes discovered had no clear homologues among the previously described genes of known function, whether from yeast or other organisms (Goffeau *et al.* 1996). The main challenge during the next stage of the yeast genome project is to elucidate the physiological role and the biochemical function of all these genes.

Considerable effort, being spent to unravel the functions of these novel genes (sometimes referred to as "functional orphans"), involves various biological approaches, for example: (i) the systematic inactivation of yeast genes by random introduction of a β-galactosidase (*lacZ*) reporter gene, generating mutant phenotypes and providing information on the level of gene expression and protein localisation (Burns *et al.*, 1994); the use of genetic footprinting to assess the phenotypic effects of Ty1 transposon insertions (Smith *et al.*, 1996); (ii) characterisation of the yeast transcriptome, providing insight into global patterns of gene expression (Velculescu *et al.*, 1997); (iii) proteome analysis through combined action of 2-D gel electrophoresis and mass spectrometry (Boucherie *et al.*, 1995; Kahn, 1995; Lamond and Mann, 1997); (iv) construction of new high-copy-number yeast vectors, designed for the conditional expression of epitope-tagged proteins *in vivo* (Cullin and Minvielle-Sebastia, 1994); or (v) *in silico* approaches (Nakai and Kanehisa, 1992; Slonimski and Brouillet, 1993; Codani *et al.*, 1998).

A joint effort of several European laboratories is under way to decipher the functions of newly discovered ORFs from yeast chromosome III as a pilot project for future studies, applicable to the whole yeast genome (Kahn, 1995). As part of this programme, we have developed a large-scale screening for the identification of biochemical and physiological functions of unknown genes by the means of systematic phenotypic analysis of individually deleted ORFs. For this purpose, some 80 ORFs of chromosome III have been deleted and a panel of some 150 different growth conditions has been developed, of which 100 are described in this chapter. In addition to the widely used standard media (e.g. discriminating between the fermentative vs respiratory growth, temperature sensitivity, sugar and nitrogen source utilisation), we have introduced a systematic inhibitor sensitivity approach. The rationale of this approach is simple. If a protein involved in a specific process is missing, the mutant cell may become more sensitive or sometimes more resistant than the wild type to the action of the inhibitor affecting this biological process itself or processes linked to it by a network of interactions. The finding of such a difference(s) under a given growth condition constitutes the first indication about the function of the mutated gene. It may be informative about the biochemical function of the deleted gene (e.g. if an increased sensitivity to a specific inhibitor is found) or it may be only indicative of the physiological role of the ORF (e.g. if a growth deficiency is found under a general stress like high temperature). Nevertheless, even in the latter case the result is useful for future studies, since it points out that the ORF in question does correspond to a real gene.

The urgent need for scaling- and speeding-up of the phenotypic testing, applicable to the continuously increasing number of available mutants to be analysed, provided by the EUROFAN project (European Functional Analysis Network), has led us to adopt microtitre-plate-based search of gene/protein functions. The aim of this chapter is to describe in detail this methodology, to illustrate it by a few examples and to discuss its advantages, drawbacks and other potential fields of applications.

◆◆◆◆◆◆ II. MATERIALS AND METHODS

A. Yeast Strains, Targeted Gene Deletions and Standard Genetic Analysis

Targeted gene deletions were carried out in either the diploid strain W303 (by *HIS3* transplacement) [*MAT***a**/*MAT*α; *ura3-1, trp1-1, ade2-1, leu2-3, 112, his3-11, 15*], the corresponding haploid strain W303-1B [*MAT*α; *ura3-1, trp1-1, ade2-1, leu2-3, 112, his3-11, 15;* (Thomas and Rothstein, 1989)] or BMA64 (by *TRP1* transplacement) [*MAT***a**/*MAT*α; *ura3-1, ade2-1, leu2-3, 112, his3-11, 15, trp1*Δ; (Baudin-Baillieu *et al*., 1997)]. Construction of ORF deletion cassettes, yeast transformation assays, PCR analysis of the transformants and Southern blot analysis were performed as described by Baudin *et al.* (1993), Coppée *et al.* (1996) and Rieger *et al.* (1997). Yeast mating, sporulation and tetrad analysis were performed as described by Rose *et al*. (1990).

B. Media Composition and Inhibitors

If not stated otherwise, inhibitors, salts, heavy metals and other chemicals were added directly to the three standard media (YPGFA, WOFA, N3FA, where FA denotes Functional Analysis) listed below under headings **001–003**. Stock solutions of the different compounds were made in acetone, ethanol, dimethyl sulfoxide (DMSO), dimethyl formamide (DMF), methanol, acetic acid and, if not further specified below, in water. Inhibitor concentrations are given below and final concentrations in the test media are listed in Table 1. Stock solutions were stored following the instructions of the suppliers. Various concentrations of solvents were assayed on wild-type strains, to exclude the possibility that solvents themselves cause growth inhibition. In some of the conditions listed below, DMSO was added to a final concentration of 3% to facilitate penetration of the corresponding inhibitor and controls (solvent alone) vs experimental (solvent + inhibitor) were compared. All chemicals were obtained from the Sigma Chemical Company (St Quentin Fallavier, FL), except for benomyl, which was a gift from E. I. DuPont (Wilmington, Del., lot: B-19501), hydroxyurea and sodium orthovanadate (Aldrich, St Quentin Fallavier, FL), maltose (Merck, Darmstadt), ferrous (II) sulfate (Serva, Heidelberg) and thiolutin (Pfizer, Groton, Conn.). All of them were of the highest available purity grade. In general, compounds were added from filter sterilised stock solutions to media cooled to about 65°C.

1. Standard media

(001) YPGFA, standard complete glucose medium: 1% yeast extract (Difco Laboratories, Detroit, USA), 1% bactopeptone (Difco), 2% glucose and 80 mg l^{-1} adenine (adenine is added in large excess in order to prevent the formation of the red pigment in ade2 strains); **(002)** WOFA, standard synthetic glucose medium: 0.67% yeast nitrogen base without amino acids (Difco), 2% glucose, supplemented with: 80 mg l^{-1} adenine, 20 mg l^{-1} uracil, 10 mg l^{-1} histidine, 60 mg l^{-1} leucine, 20 mg l^{-1} tryptophan; **(003)** N3FA, standard glycerol medium: 1% yeast extract, 1% bactopeptone, 2% glycerol, 0.05 M sodium phosphate (pH 6.2, 100 ml l^{-1}) and 80 mg l^{-1} adenine. Media were solidified by adding 2% (Petri dishes) or 0.7% Bacto-Agar (Difco) to 96-well microtitre plates (Nunc Intermed, Polylabo, Paris).

2. Salts and heavy metals

The following compounds were added to YPGFA before autoclaving (for final concentrations see Table 1): KCl, NaCl, $MgCl_2$, $MgSO_4$, NH_4Cl, $SrCl_2$.

(004) 001 + $BaCl_2$ [1 M]; **(005)** 001 + $FeCl_2$ [0.3 M]; **(006)** 001 + $FeCl_3$ [0.2 M]; **(007)** 001 + $FeSO_4$ [0.2 M]; **(008)** 001 + $CaCl_2$ [5 M]; **(009)** 001 + $CdCl_2$ [1 mM]; **(010)** 001 + CsCl [3 M]; **(011)** 001 + $CoCl_2$ [0.3 M]; **(012)** 001 + $CuSO_4$ [0.5 M]; **(013)** 001 + $NiCl_2$ [0.3 M]; **(014)** 001 + $HgCl_2$ [0.2 M]; **(015)** 001 + KCl; **(016)** 001 + NaCl; **(017)** 001 + $MgCl_2$; **(018)** 001 + $MgSO_4$; **(019)** 001 + NH_4Cl; **(020)** 001 + RbCl [4 M]; **(021)** 001 + $SrCl_2$; **(022)** 001 + LiCl [5 M]; **(023)** 001 + $MnCl_2$ [0.1 M]; **(024)** 001 + $ZnCl_2$ [0.1 M].

3. Inhibitors

(025) 002 + hydroxyurea [100 mg ml^{-1}]; **(026)** 002 + phenylethanol [100 mg ml^{-1} in ethanol]; **(027)** 003 + nalidixic acid [10 mg ml^{-1} in 1 N NaOH] + 3% DMSO; **(028)** 002 + actinomycin D [0.8 mg ml^{-1} in ethanol] + 3% DMSO; **(029)** 002 + 8-hydroxyquinoline [1 mg ml^{-1} in ethanol]; **(030)** 002 + cycloheximide [0.1 mg ml^{-1}]; **(031)** 002 + anisomycin [2 mg ml^{-1} in ethanol]; **(032)** 002 (supplemented with 5 μg ml^{-1} uracil) + 6-azauracil [3.5 mg ml^{-1}]; **(033)** 001 + protamine sulfate [10 mg ml^{-1}]; **(034)** 001 + chlorambucil [0.3 M in cold acetone]; **(035)** 003 + antimycin A [1 μg ml^{-1} in cold acetone]; **(036)** 003 + chloramphenicol [100 mg ml^{-1} in ethanol]; **(037)** 003 + erythromycin [100 mg ml^{-1} in acetone]; **(038)** 001 + benomyl [5 mg ml^{-1} in DMSO]; **(039)** 001 + caffeine [5%]; **(040)** 003 + sodium orthovanadate [0.05 M in 50 mM KOH]; **(041)** 002 + sodium fluoride [1 M]; **(042)**

Table 1. The list of the first 100 growth media for the phenotypic analysis of genes of unknown function from yeast chromosome III. Numbers refer to the preparation of the corresponding media as outlined in Section II, Materials and methods. Concentrations given in the table were set up with the corresponding haploid wild-type strains (W303-1B [*MAT*α] and W303-1B/A [MAT**a**]). Literature quotations are non-exhaustive and indicative only

	Compound	Final concentration	Function and/or target and mode of function of inhibitors
001	YPGFA	—	standard complete glucose medium
002	WOFA	—	standard synthetic glucose medium
003	N3FA	—	standard complete glycerol medium (respiratory growth)
004	$BaCl_2$	50 mM	ion-transport (Borst-Pauwels, 1981)
005	$FeCl_2$	23 mM	transport; toxicity through generation of hydroxyl radicals (Kosman, 1994; Georgatsou and Alexandraki, 1994)
006	$FeCl_3$	8.5 mM	transport; toxicity through generation of hydroxyl radicals (Kosman, 1994; Georgatsou and Alexandraki, 1994)
007	$FeSO_4$	23 mM	transport; toxicity through generation of hydroxyl radicals (Kosman, 1994; Georgatsou and Alexandraki, 1994)
008	$CaCl_2$	0.5 M	ion-transport (Borst-Pauwels, 1981), cell cycle regulation (Iida *et al.*, 1990)
009	$CdCl_2$	40–50 μM	growth inhibition, transport (Conklin *et al.*, 1993; Romandini *et al.*, 1992)
010	CsCl	0.1 M	transport (Bossemeyer *et al.*, 1989) growth inhibition, K^+ replacement (Perkins and Gadd, 1993)
011	$CoCl_2$	750 μM	transport, growth inhibition, resistance (Conklin *et al.*, 1994; Kosman, 1994)
012	$CuSO_4$	5–6 mM	transport, growth inhibition (Conklin *et al.*, 1993; Dancis *et al.*, 1994; Romandini *et al.*, 1992)
013	$NiCl_2$	850 μM	transport, potential inhibitor of the uptake of other metal ions (Conklin *et al.*, 1993; Kosman, 1994)
014	$HgCl_2$	230–250 μM	growth inhibition (Farrell *et al.*, 1993)
015	KCl	1.3 M	salt tolerance, ion transport (Borst-Pauwels, 1981; Gaxiola *et al.*, 1992)

Table 1. *Continued*

	Compound	Final concentration	Function and/or target and mode of function of inhibitors
016	NaCl	1.3 M	salt tolerance, ion transport (Borst-Pauwels, 1981; Gaxiola *et al.*, 1992)
017	$MgCl_2$	0.5/0.7 M	salt tolerance, ion transport (Borst-Pauwels, 1981)
018	$MgSO_4$	0.4/0.6 M	salt tolerance, ion transport (Borst-Pauwels, 1981)
019	NH_4Cl	0.7/1 M	ion-transport (Borst-Pauwels, 1981)
020	RbCl	0.2 M	ion-transport (Borst-Pauwels, 1981)
021	$SrCl_2$	0.5 M	ion-transport (Borst-Pauwels, 1981)
022	LiCl	0.15–0.175 M	transport, growth inhibition (Conklin *et al.*, 1993; Perkins and Gadd, 1993)
023	$MnCl_2$	4 mM	electrophilic prosthetic group in several enzymes, transport (Conklin *et al.*, 1993; Kosman, 1994), RNA processing
024	$ZnCl_2$	4–5 mM	electrophilic prosthetic group in several enzymes, transport (Conklin *et al.*, 1993, 1994; Kosman, 1994)
025	hydroxyurea	6 mg ml^{-1}	inhibitor of DNA synthesis (Schindler and Davies, 1975)
026	phenylethanol	2 mg ml^{-1}	inhibitor of DNA synthesis (Schindler and Davies, 1975)
027	nalidixic acid	200 μg ml^{-1}	inhibits DNA synthesis (Schindler and Davies, 1975)
028	actinomycin D	45 μg ml^{-1}	inhibitor of RNA synthesis (Schindler and Davies, 1975)
029	8-hydroxyquinoline	26 μg ml^{-1}	chelating agent, RNA synthesis inhibitor (Schindler and Davies, 1975)
030	cycloheximide	0.2/0.3 μg ml^{-1}	protein synthesis inhibitor (Tuite, 1989)
031	anisomycin	50 μg ml^{-1}	inhibitor of protein synthesis (Schindler and Davies, 1975)
032	6-azauracil	350 μg ml^{-1}	growth inhibitor, inhibitor of GTP synthesis (Exinger and Lacroute, 1992)
033	protamine sulfate	750 μg ml^{-1}	acts on plasma membrane ATPase
034	chlorambucil	2/3 mM	alkylation agent, mutagen, acts on DNA repair processes (Ruhland and Brendel, 1979)
035	antimycin A	0.0025 μg ml^{-1}	inhibitor of mitochondrial respiration chain (Slater, 1973)
036	chloramphenicol	2 mg ml^{-1}	inhibitor of the mitochondrial peptidyl transferase (Meyers *et al.*, 1992)

037	erythromycin	200 μg ml^{-1}	inhibitor of mitochondrial protein synthesis (Treinin and Simchen, 1993)
038	benomyl	25/40 μg ml^{-1}	anti-microtubule drug (Li and Murray, 1991)
039	caffeine	0.15–0.2%	inhibitor of cAMP-phosphodiesterases (Beach *et al.*,1985; Parsons *et al.*, 1988)
040	sodiumorthovanadate	3 mM	inhibition of mitochondrial H^+-and plasma membrane Na^+-K^+-ATPases (Henderson *et al.*, 1989), vanadate-resistant mutants are defective in protein glycosylation (Ballou *et al.*, 1991)
041	sodium fluoride	5 mM	inhibits various phosphatases (Farkas, 1989)
042	1,10-phenanthroline	30–35 μg ml^{-1}	chelating agent, causes Zn^{2+} and/or Fe^{2+} deprivation (Bossier *et al.*, 1993; Oliver and Warmington, 1989)
043	cerulenin	0.5 μg ml^{-1}	inhibits biosynthesis of fatty acids (Omura, 1981)
044	2,2-dipyridyl	50 μg ml^{-1}	chelator of divalent cations
045	aurintricarboxylic acid	100 μM	inhibitor of protein synthesis (Battaner and Vazquez, 1971)
046	staurosporine	3.5 μg ml^{-1}	specific inhibitor of protein kinase C (Toda *et al.*, 1991; Yoshida *et al.*,1992)
047	colchicine	2 mg ml^{-1}	disassembly of microtubules (Manfredi and Horwitz, 1984)
048	trifluoperazine	500 μM	calcium channel blocker (Bruno and Slate, 1990), inhibitor of calcium binding proteins
049	verapamil	100 μg ml^{-1}	calcium channel blocker (Bruno and Slate, 1990), phospholipid interacting drug
050	cinnarizine	100 μg ml^{-1}	inhibits uptake of cations (Janis *et al.*, 1987)
051	tunicamycin	2.5 μg ml^{-1}	blocks incorporation of mannose into the *N*-glycans of glycoproteins (Sipos *et al.*, 1994)
052	griseofulvin	100 μg ml^{-1}	disorganises microtubules (Manfredi and Horwitz, 1984)
053	PMSF	4–5 mM	inhibitor of serine proteases
054	L-ethionine	1 μg ml^{-1}	inhibits methylation, *S*-adenosyl methionine formation
055	paromomycin sulfate	2 mg ml^{-1}	protein synthesis inhibitor (Tuite, 1989)
056	5-azacytidine	100 μg ml^{-1}	inhibits several bacterial DNA (cytosine-5) methylases (Friedman, 1982)

Table 1. *Continued*

	Compound	Final concentration	Function and/or target and mode of function of inhibitors
057	brefeldin A	100 μg ml^{-1}	blocks protein transport out of the golgi apparatus (Jackson and Képès, 1994)
058	nocodazole	100 μg ml^{-1}	antimicrotubule drug (Manfredi and Horwitz, 1984; Torres *et al.*, 1991)
059	thiolutin	2–9 μg ml^{-1}	inhibitor of all 3 yeast RNA polymerases (Oliver and Warmington, 1989)
060	CCCP	2/3 μM	uncoupler of oxidative phosphorylation, protonophor (Xu and Shields, 1993)
061	oligomycin	0.2/0.3 μg ml^{-1}	inhibits mitochondrial ATPase (Treinin and Simchen, 1993)
062	neomycin sulfate	0.5–1 mg ml^{-1}	inhibits mitochondrial and cytoplasmic protein synthesis (Dujon, 1981)
063	emetine	2 mg ml^{-1}	protein synthesis inhibitor (Battaner and Vazquez, 1971)
064	acetylsalicylic acid	0.4–0.5 mg ml^{-1}	interferes with heme biosynthesis
065	fluorescent brightener 28	2 mg ml^{-1}	amplifies the effect of cell wall mutations (Ram *et al.*, 1994)
066	PCMB	0.3 mM	non-specific inhibitor of metalloproteases
067	nystatin	4–5 μg ml^{-1}	membrane-active antifungal agent, binds to sterols (Gennis, 1989)
068	2,4-dinitrophenol	0.4 mM	protonophore
069	tetraethylammonium chloride	0.1 M	K^+ channel blocker (Anderson *et al.*, 1992)
070	3-amino-1,2,4-triazole	2.5 mg ml^{-1}	catalase inhibitor (Van der Leij *et al.*, 1992)
071	diltiazem hydrochloride	2 mg ml^{-1}	calcium channel blocker (Bruno and Slate, 1990)
072	EDTA	1 mg ml^{-1}	metal-ion chelating agent, non-specific inhibitor of metalloproteases
073	ethanol	10–15%	ethanol tolerance
074	formamide	2.5–3%	formamide sensitivity as a conditional phenotype (Aguilera, 1994)
075	dimethylformamide	2.5–3%	
076	diamide	1.6 mM	thiol oxidising agent, oxidative stress (Kuge and Jones, 1994)
077	H_2O_2	1–2.5 mM	oxidative stress (Kuge and Jones, 1994)

078	L-canavanine	30 μg ml^{-1}	inhibits arginine permease
079	2-deoxy-D-glucose	200 μg ml^{-1}	causes repression of glucose-repressible genes but is not used as a carbon source (Neigeborn and Carlson, 1987)
080	sorbitol	1.8 M	salt tolerance (Gaxiola *et al.*, 1992)
081	potassium acetate	3%	carbon source (Fernandez *et al.*, 1994)
082	ethanol	3%	carbon source (Fernandez *et al.*, 1994)
083	maltose	2%	carbon source (Fernandez *et al.*, 1994)
084	galactose	2%	carbon source (Fernandez *et al.*, 1994)
085	sucrose	2%	carbon source (Fernandez *et al.*, 1994)
086	raffinose	2%	carbon source (Fernandez *et al.*, 1994)
087	melibiose	2%	carbon source (Fernandez *et al.*, 1994)
088	fructose	2%	carbon source (Fernandez *et al.*, 1994)
089	lactate	2%	carbon source (Fernandez *et al.*, 1994)
090	oleic acid	0.25%	induction of peroxisome proliferation, carbon source (Van der Leij *et al.*, 1992)
091	lauric acid	0.05%	carbon source (Van der Leij *et al.*, 1992)
092	proline	1 mg ml^{-1}	nitrogen source
093	allantoin	1 mg ml^{-1}	nitrogen source
094	glutamic acid	1 mg ml^{-1}	nitrogen source
095	L -glutamine	1 mg ml^{-1}	nitrogen source
096	NH_4Cl	1 mg ml^{-1}	nitrogen source
097	L -ornithine	1 mg ml^{-1}	nitrogen source
098	L -serine	1 mg ml^{-1}	nitrogen source (Petersen *et al.*, 1988)
099	L -threonine	1 mg ml^{-1}	nitrogen source (Petersen *et al.*, 1988)
100	urea	1 mg ml^{-1}	nitrogen source

002 + 1,10 phenanthroline [10 mg ml^{-1} in ethanol]; (**043**) 001/002 + cerulenin [1 mg ml^{-1} in ethanol]; (**044**) 002 + 2,2 dipyridyl [10 mg ml^{-1}]; (**045**) 002 + aurintricarboxylic acid [2 mM in ethanol]; (**046**) 001 + staurosporine (0.5 mg ml^{-1} in DMSO]; (**047**) 002 + colchicine [100 mg ml^{-1} in ethanol]; (**048**) 002 + trifluoperazine [0.01 M]; (**049**) 002 + verapamil hydrochloride [2 mg ml^{-1} in ethanol]; (**050**) 002 + cinnarizine [1 mg ml^{-1} in ethanol]; (**051**) 002 + tunicamycin [1.1 mg ml^{-1} in 0.001 M NaOH]; (**052**) 002 + griseofulvin [10 mg ml^{-1} in DMF]; (**053**) 002 + phenylmethylsulfonyl fluoride (PMSF) [0.1 M in methanol]; (**054**) 002 + L-ethionine [1 mg ml^{-1}]; (**055**) 002 + paromomycin sulfate [100 mg ml^{-1}]; (**056**) 002 + 5-azacytidine [2.5 mg ml^{-1}]; (**057**) 001 + brefeldin A [5 mg ml^{-1} in methanol]; (**058**) 001 + nocodazole [2.5 mg ml^{-1}]; (**059**) 002 + thiolutin [0.2 mg ml^{-1} in DMSO]; (**060**) 003 + carbonyl-cyanide *m*-chlorophenylhydrazone (CCCP) [10 mM in ethanol]; (**061**) 003 + oligomycin [2 mg ml^{-1} in ethanol]; (**062**) 003 + neomycin sulfate [5 mg ml^{-1}]; (**063**) 002 + emetine [20 mg ml^{-1} in ethanol]; (**064**) 002 + acetylsalicylic acid [100 mg ml^{-1} in ethanol]; (**065**) 001 + fluorescent brightener 28 [20 mg ml^{-1}]; (**066**) 001 + *p*-chloromercuribenzoic acid (PCMB) [10 mM in DMSO]; (**067**) 001 + nystatin [1 mg ml^{-1}]; (**068**) 003 + 2,4-dinitrophenol [20 mM in acetone]; (**069**) 001 + tetraethylammonium chloride [1 M]; (**070**) 002 + 3-amino-1,2,4-triazole [100 mg ml^{-1}]; (**071**) 002 + diltiazem hydrochloride [50 mg ml^{-1}]; (**072**) 001 + ethylenediaminetetraacetic acid (EDTA) [10 mg ml^{-1}]; (**073**) 001 + ethanol [100%]; (**074**) 001 + formamide [100%]; (**075**) 001 + dimethylformamide [100%]; (**076**) 001/002 + diamide [50 mM]; (**077**) 001 + H_2O_2 [30%]; (**078**) YCBFA + L-canavanine [2 mg ml^{-1}]; (**079**) YPFA + 2-deoxy-D-glucose [0.2 mg ml^{-1} YPFA medium containing 2% sucrose or 2% galactose]; (**080**) 001 + 1.8 M sorbitol.

4. Carbon sources

Standard complete medium without glucose (YPFA) or standard synthetic medium without glucose (WOFA) was supplemented with the corresponding carbon source: (**081**) YPFA/WOFA + 3% potassium acetate; (**082**) YPFA/WOFA + 3% ethanol; (**083**) YPFA/WOFA + 2% maltose; (**084**) YPFA/WOFA + 2% galactose; (**085**) YPFA/WOFA + 2% sucrose; (**086**) YPFA/WOFA + 2% raffinose; (**087**) YPFA/WOFA + 2% melibiose; (**088**) YPFA/WOFA + 2% fructose; (**089**) YPFA/WOFA + 2% lactate; (**090**) oleic acid (0.67% yeast nitrogen base (Difco), 2.5% Bacto-Agar (Difco), 0.05% yeast extract (Difco), 0.25% oleic acid [10% oleic acid, 10% Tween 80 were mixed with 70 ml of water prior to addition of 0.7 g NaOH in 10 ml water], growth factors as in 002); (**091**) lauric acid (0.67% yeast nitrogen base (Difco), 2.5% Bacto-Agar (Difco), 0.05% yeast extract (Difco), 0.05% lauric acid from a stock containing 1% lauric acid and 8.3% Tween 40, growth factors as in 002, the pH of the medium was adjusted to pH 6).

5. Nitrogen sources

Under the conditions listed below the standard medium is YCBFA (1.17% yeast carbon base (Difco), 0.1% K_2HPO_4, 2% glucose, 20 mg l^{-1} adenine,

20 mg l^{-1} uracil, 10 mg l^{-1} histidine, 60 mg l^{-1} leucine, 20 mg l^{-1} tryptophan) supplemented with the following compounds as sole nitrogen source: (**092**) YCBFA + proline [1 mg ml^{-1}]; (**093**) YCBFA + allantoin [1 mg ml^{-1}]; (**094**) YCBFA + glutamic acid [1 mg ml^{-1}]; (**095**) YCBFA + L-glutamine [1 mg ml^{-1}]; (**096**) YCBFA + NH_4Cl [1 mg ml^{-1}]; (**097**) YCBFA + L-ornithine [1 mg ml^{-1}]; (**098**) YCBFA + L-serine [1 mg ml^{-1}]; (**099**) YCBFA + L-threonine [1 mg ml^{-1}]; (**100**) YCBFA + urea [1 mg ml^{-1}].

C. General Culture Conditions

(i) Three growth temperatures (16, 28, 36°C); (ii) plate assay for heat shock sensitivity: fresh cells grown overnight in liquid YPGFA medium at 28°C were serially diluted to 1:100 and 1:1000 in Ringer and 20 µl of the corresponding mutant- or wild type cell suspensions were spotted on plates containing media **001–003.** The plates were sealed with Parafilm, floated in a water bath and incubated for 60 min at 55°C. Then, the plates were cooled to room temperature and incubated at 28°C for 3 or 4 days until heat shock sensitivity was scored; (iii) osmotic lability: about 5×10^7 cells from an overnight culture at 28°C were arranged in a cluster tube eight-strip rack (Costar, Polylabo, Paris), washed twice in sterilised water and shaken at 28°C for up to 10 days. Viability was checked by spotting 1 : 100 and 1:1000 in water diluted aliquots (5 µl) of all cultures on YPGFA and WOFA containing microtitre plates; (iv) pH: concentrated YPGFA (90% of final volume) was mixed at 60°C with filter-sterilised 10 × acetate buffer (1 M) of pH in the range 2.41 to 5.51.

D. Establishing the Range of Inhibitor Concentrations for the Reference Strain

The first step consisted of establishing the threshold concentration (or a range of concentrations) for the reference strain. The threshold concentration should be not too high in order to allow the growth of the reference strain and easy discrimination of hypersensitive mutants and not too low in order to detect significant increase in resistance in other mutants. To this end, standard or special media (5 ml) were supplemented at about 65°C with compounds to be tested and poured in Petri dishes (Ø 55 mm). Reference strains [W303-1B (*MATα*) and W303-1B/A (*MAT**a***, isogenic to the previous and obtained by mating type switch)] were pre-grown overnight in liquid YPGFA at 28°C. Dilutions made in Ringer were spotted (5 µl of 1:100 and 1:1000) on plates and grown up to 7 days at 28°C and 36°C in the presence or absence of the desired drug. Following incubation at these temperatures, growth was assessed visually at least every 24 h.

E. Phenotypic Tests in Microtitre Plates

In general, standard media (50 ml) were sterilised in Erlenmeyer flasks and stored at 4°C. Solid media were liquefied by heating (85°C) in a covered

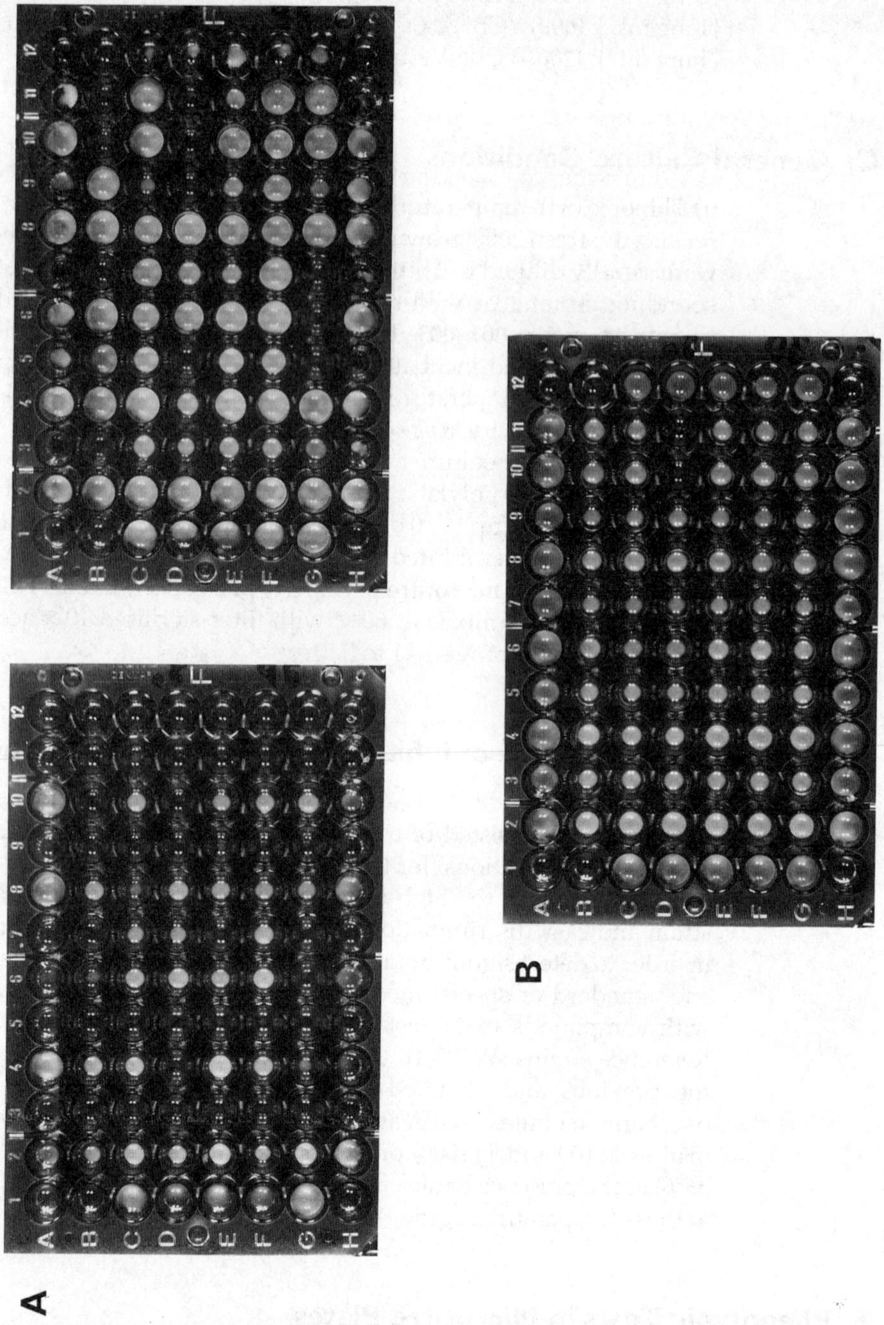

water bath (Salvig, Reussbühl, CH). After cooling to about 65°C, inhibitors were added. Solutions were then transferred to a multipipette adapted disposer, out of which, using automatic multichannel pipettes, they were filled in flat bottomed 96-well microtitre plates at about 230 μl per well. Control and deleted strains were pre-grown overnight in fully aerated, shaken liquid YPGFA at 28°C to early stationary phase (*ca.* 2–4 × 10^8 cells per ml). Aliquots (0.5 ml) of cultures were gridded in cluster tube eight-strip racks, serving as a master plate, and subsequently serially diluted in Ringer. Twenty microlitres of 1 : 100 and 1 : 10000 diluted cell suspensions, corresponding to about 2–4 × 10^4 and 2–4 × 10^2 cells, respectively, were inoculated into the wells and the microtitre plate placed on a shaker for 10 seconds in order to cover uniformly the agar surface. Plates were then incubated at 16°C, 28°C and 36°C for up to 12 days. From the first day of incubation, growth of the mutant strains was scored visually, either directly on the plate or later on photographs (see Figures 1 and 2), against the growth of the corresponding control strains.

◆◆◆◆◆◆ III. RESULTS AND DISCUSSION

A. Systematic Phenotype Screening

The aim of the present study is to describe an efficient methodology applicable to a large-scale phenotypic analysis of the yeast *S. cerevisiae* genome. The search for phenotypic consequences resulting from the inactivation of

Figure 1. Large-scale phenotypic tests in microtitre plates for increased drug sensitivity. Representative examples from the screening of deleted ORFs of unknown function, from yeast chromosome III. Preparation of media and cell suspensions was done as outlined in Section II, Materials and methods. Microtitre plates with control (in triplicate) and deleted strains (42 strains corresponding to deletions in 26 different ORFs, and representing either independent isolates of the same deletions or the two mating types, *MATα* and *MAT*a, carrying the same deletion) were incubated for various periods of time at 28°C and 36°C. Strains were arranged in the plates as follows: vertical rows with numbers 2, 4, 6, 8, 10 and 1 contain high cell inoculum (*ca.* 2 × 10^4 cells per well), vertical rows with numbers 3, 5, 7, 9, 11 and 12 contain low cell inoculum (*ca.* 2 × 10^2 cells per well). Each strain is inoculated twice (one strain in wells A2 and A3, second in B2 and B3, third in C2 and C3,..., C1 and C12,..., G1 and G12). The corner wells and B1, B12 are not used and the control wild-type strain W303-1B occupies positions D2, D3, E2, E3 and F2, F3. (A) Screening for caffeine sensitivity. Evolution of growth photographed after 4 days (left) and 11 days (right) after incubation at 28°C in the presence of 0.2% caffeine (growth medium: 039). Phenotypic class 0: no growth even after longer incubation with both inocula (e.g. strains: B10, B11; D10, D11; G6, G7). Phenotypic class 1: no growth at 4 days, diminished growth with high inoculum and no growth with low inoculum after longer incubation (e.g. strains: A6, A7 and D4, D5). Phenotypic class 2: growth equal to the control with high inoculum after longer incubation, poor or no growth with low inoculum (e.g. strains: C8, C9 and H4, H5). Phenotypic class 3: growth with both inocula equal to the control (e.g. strains: C4, C5 and F6, F7). (B) Screening for sensitivity to hydroxyurea (6 mg ml^{-1}; growth medium: 025). The photograph was taken after 4 days of incubation at 28°C. Strain in D10, D11 belongs to the phenotypic class 0.

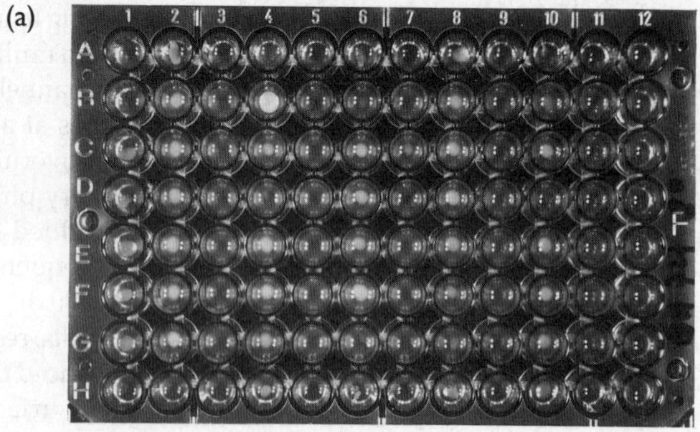

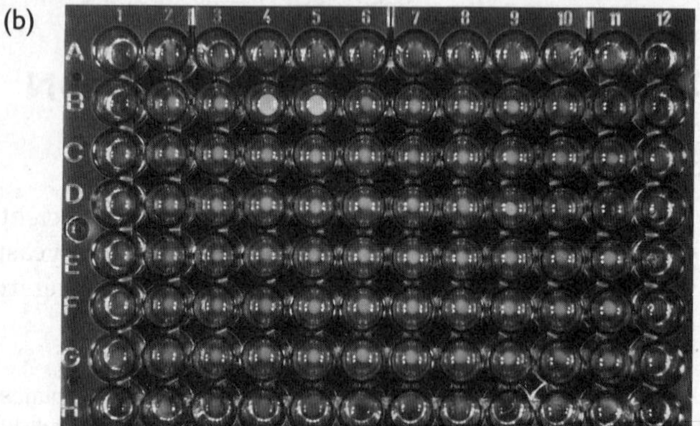

Figure 2. Large-scale phenotypic tests in microtitre plates for increased drug resistance. Conditions for incubation and arrangement of strains are as depicted in Figure 1. The control wild-type strain W303-1B occupies positions D2, D3, E2, E3 and F2, F3. Growth was assessed in the presence of 2.5 μg ml^{-1} tunicamycin (medium: 051) or 2 mg ml^{-1} phenylethanol (medium: 026). Photographs were taken after 6 days (tunicamycin, part a) and 5 days (phenylethanol, part b) of incubation at 28°C. Only one mutant strain shows a strong resistance to tunicamycin (B4, high inoculum) and phenylethanol (B4 and B5, high and low inoculum). Note that the control strains grow poorly under these conditions.

individual genes is the first step which follows logically the determination of the complete sequence of the yeast genome, necessary for the understanding of the biology of this organism. This search should fulfil simultaneously two criteria: (i) it should be as broad, exhaustive and unbiased as possible; (ii) it should be practical, i.e. easily reproducible, applicable as a routine and not too time consuming. Apparently, these two criteria are contradictory, since the number of imaginable growth conditions is enormous and therefore screening for all of them is impossible. However, some several hundred growth conditions are sufficient to cover in an initial screening a large fraction of biochemical, developmental, regulatory and signalling pathways of the yeast cell. Once a clear mutant phenotype has been revealed, a discrete inhibited step in a pathway may be further characterised, for example, by the use of analogues or unrelated compounds acting in the same general process (e.g. testing for respiration deficiency in the presence of acetate, ethanol, glycerol and lactate). On the basis of these findings, the first 100 growth conditions, covering an important part of the yeast biology, were selected (Table1).

Given the increasing number of mutants to be analysed and the potential applications in screening of chemical compounds (hunt for interesting new drug candidates), we adopted microtitre plate technology to search for phenotypes. The advantages and drawbacks of this system can be summarised as follows: (i) easy to handle large numbers of strains and conditions; smaller volume for storage incubation; ii) straightforward to score phenotypic differences (see Figures 1 and 2); (iii) less expensive, especially for costly or rare chemicals (in our tests, total volume of 96-well microtitre plate is less than 25 ml, while a single Petri dish requires 20–25 ml and allows analysis of only one-fifth of strains by the drop-out technique; (iv) absence of cross-feeding and cross-diffusion between individual drop-out cultures (e.g. diffusion of secreted enzymes or metabolites); (v) analysis on solid or in liquid media; (vi) simultaneous analysis of at least 60 strains under optimal growth conditions in one microtitre plate; (vii) quick, simple and provides well-reproducible results; (viii) possibility of automation (see Future developments, Section III.B below).

Some of the critical points of this experimental approach concern: (i) in the well, the agar surface is concave and smaller than the surface of a drop out deposit on a flat surface of a Petri dish. Therefore, analysis of colonies grown from individual cells (their number, shape and morphological heterogeneity) is more easily analysed on Petri dishes than on wells; (ii) optimal growth conditions are available in all wells of the plate, except for the outer ring of wells, where growth differences may result from an accelerated evaporation (corner wells should never be used because of this phenomenon).

For all phenotypic tests, "calibration" of growth conditions in respect to a reference strain is required. The analysed mutants were derived from three different "wild-type" genetic backgrounds (W303, FY1679, CEN.PK2). Depending on the genetic background, important differences in sensitivity to a given drug were observed. This was particularly true for the CEN.PK2 strain. For this strain, some 30% of the tested growth conditions turned out to be unsuitable for phenotypic analysis since inhibitor concentrations were either below or over the threshold determined for W303. Otherwise, in most cases growth differences between W303 and FY1679 were negligible, except for respiratory media (Rieger *et al.*, 1997).

All these points have been taken into consideration in order to obtain reliable and informative results. Some representative examples from this large-scale screening are shown in Figures 1 and 2. As illustrated for the screening in the presence of caffeine and hydroxyurea (Figure 1A, B), some mutants display complete inhibition of growth which can be easily detected. Furthermore, examination of growth as a function of time can even detect subtle variations in growth rates (Figure 1A). These mutants, with either a reduced growth for both inocula or no growth of the low inoculum, are only indicative and classified as suggestive phenotypes (Figure 3B). Growth conditions have also been adjusted to screen for drug-resistant mutants (Figure 2). Other examples can be found in Rieger *et al.*, (1997).

In general, the notion of the "function" of a gene is of necessity ambiguous. It should be defined according to various levels of analysis: physiological role, participation in cellular processes and biochemical pathways, underlying molecular mechanisms, etc. These various "functions" can be deduced either from experiments (*in vivo* or *in vitro* approaches) or from

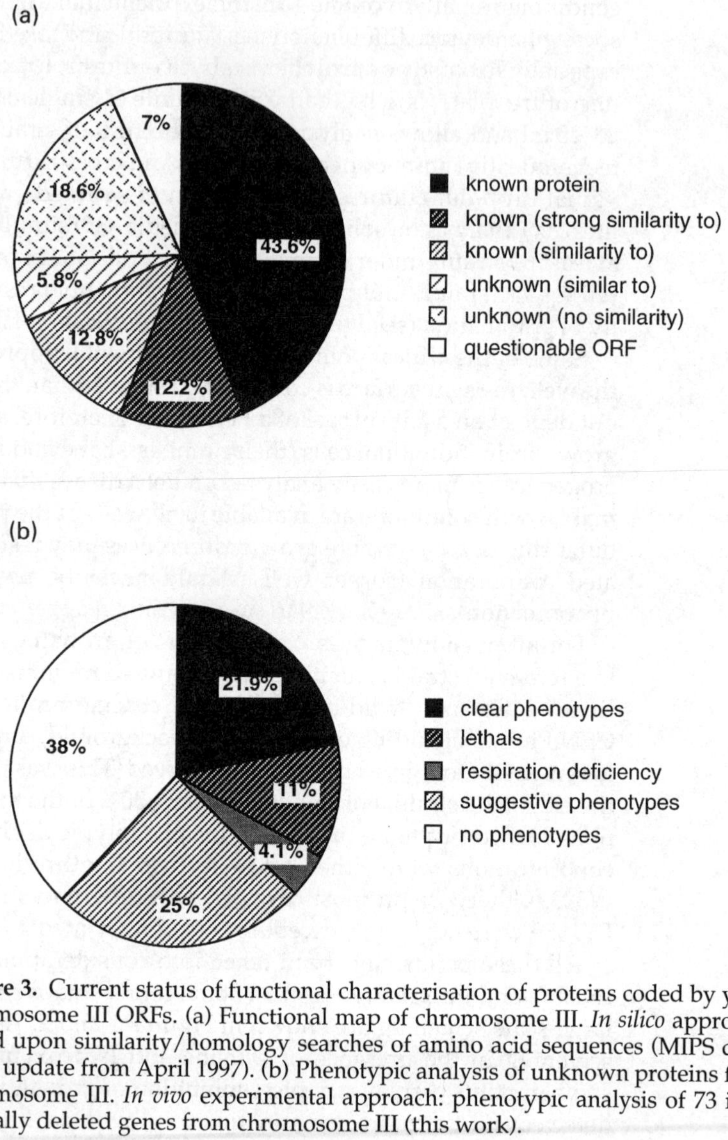

Figure 3. Current status of functional characterisation of proteins coded by yeast chromosome III ORFs. (a) Functional map of chromosome III. *In silico* approach: based upon similarity/homology searches of amino acid sequences (MIPS database, update from April 1997). (b) Phenotypic analysis of unknown proteins from chromosome III. *In vivo* experimental approach: phenotypic analysis of 73 individually deleted genes from chromosome III (this work).

similarity/homology comparisons at the sequence [complete proteins or fragments (Expressed Sequence Tags, ESTs)] level (*in-silico* approach). The latter approach is the most frequently used. According to the MIPS database, 56% of the ORFs belong to classes 1 and 2 of known proteins or display strong similarity to known proteins (higher than one-third of FASTA self-score) whereas the remaining 44% are functionally still uncharacterised (Figure 3A), belonging to ORF classes 3–6 (3, similarity to known protein; 4, similar to unknown protein; 5, no similarity; 6, questionable ORF). Of 73 genes on yeast chromosome III, belonging almost exclusively to ORF classes 3–6, tested in about 60 different growth conditions, 62% showed some phenotype, of which 37% were clear phenotypes (e.g. no growth in the presence of an inhibitor or a non-fermentable substrate, lethals) whereas no phenotype was found for 38% of the analysed genes (Figure 3B). In conclusion, the experimental approach applied here allows the detection of phenotypes precisely for those ORFs for which no indications about their biological/physiological role are available.

Nevertheless, one has to keep in mind that a phenotype is only the starting point of a functional analysis of a given gene. Its interpretation relies, on the one hand, on potentially significant sequence similarities with known genes and, on the other hand, on our knowledge about the cellular target(s) and mode(s) of action of inhibitors. For example, complete growth inhibition in the presence of sodium fluoride, a phosphatase inhibitor, implicates a relatively discrete function in the cell, whereas no growth on caffeine leaves us with a panoply of possibly affected cellular processes, including DNA repair and recombination, intracellular calcium homeostasis and cell cycle progression. At the level of phenotypic tests, we cannot differentiate between the primary lesion caused by the deletion and a secondary effect or a general unhealthy state of the cell. This must be established by further more detailed studies.

But to this end an important step towards understanding of function has been made – the gene is now accessible for genetic/biochemical analysis. A clear and stringent phenotype could be used to search for genetic interactions via isolation of multicopy and extragenic suppressors, testing of interaction between mutations with similar phenotypes, which would provide further information about the function of the studied ORF. In concert with different but complementary approaches like transcript analysis, 2-D gel electrophoresis of proteins and 2-hybrid analysis to mention only some, a coherent picture of the role of various novel genes in integrated cellular processes should emerge.

B. Future Developments

The results presented so far were obtained with the goal of uncovering clear phenotypes which should lead to a better understanding of gene function. A supplementary, potential use of this *in vivo* screening system is to identify new targets for chemical compounds, coming either from yeast or from other organisms. The utility of yeast as a model organism for high-throughput screening (HTS) is therefore two-fold – combining the hunt for gene function and drug research. For this purpose three elements are required: (i) automation of the microtitre-plate-based *in vivo* screening system (robotics workstation, computerised system for data acquisition, collection and management); (ii) suitable arrayed libraries of chemical compounds; (iii) suitable standardised mutant collection. The attractiveness of this approach relies on the advantages of yeast and "small" chemical compounds (Table 2). There are, of course, certain disadvantages to this system as well. For example, yeast lacks some higher order functions present in Metazoa. In addition, "small" chemical compounds may lack the target specificity which theoretically can be obtained with macromolecules as therapeutic agents.

This table should not be regarded as an exhaustive list of the advantages of yeast as a model organism for HTS of chemical compounds but rather as a rationale of experimentation for future developments. We would like to insist on a point which seems to us essential in this rationale, i.e. the *in vivo* approach, which has distinct advantages over pure *in vitro* assays. If a phenotype is observed as a result of interaction between a gene mutation and a chemical compound, then there must be a biological process underlying it. In the context of *in vivo* assays, higher eukaryotes have well-known drawbacks in comparison to *Saccharomyces cerevisiae*: mammalian cells are difficult to manipulate genetically, their culture is expensive and not adapted for satisfactory propagation in HTS systems. In addition, the yeast system can be used as a tool to mimic a specific human physiological process, for example signalling through human G-protein-coupled receptors (GPCRs) or reconstitution of mammalian ion channels (for a review see Broach and Thorner, 1996).

The presented *in vivo* analysis system, although being basically a classical approach, can be adapted to HTS and thus provide a tool for the discovery of new "small"-molecule drugs. This approach would complement other strategies, including *in vitro* HTS (Broach and Thorner, 1996) against defined targets (e.g. enzymes, cloned receptors), bioinformatics, combinatorial chemistry (Hogan, 1996 for review; Verdine, 1996) and the development of macromolecular, mechanism-based therapeutic agents (e.g. oligonucleotides, genes/gene fragments, recombinant proteins).

In conclusion, such a project might be of considerable therapeutic as well as molecular interest, satisfying at the same time fundamental and applied research goals and will strengthen the role of *S. cerevisiae* as a model organism for future studies.

Acknowledgements

This work was supported by grants BIO/2.CT93.0022 (Experimental pilot study for European cooperation on gene function search in *S. cerevisiae*) from the E.C and 92H00882 from the Ministère de la Recherche. K.-J. R. and J.-Y. C. received a fellowship from the EC (ERBCHBGGCT920087) and G. O. and A. K. had fellowships from the Jumelage Franco-Polonais du CNRS and the Reseaux de Formation-Recherche from the Ministère de la Recherche. We are grateful to our colleagues J. P. di Rago,

Table 2. Yeast – a model organism for *in vivo* high-throughput screening (HTS) of chemical compounds

Advantages of yeast	Advantages of chemical compounds
• well-known eukaryote (complete genome with 6000 genes) • many homologies to human genes • large mutant collection • important industrial microbe • no ethical concerns • automation of large-scale phenotypic analysis • characterisation of genes/proteins (e.g. from mammals, plants) by heterologous complementation of yeast mutants • unicellular, grows on chemically defined media which allows "complete" control of its physiology • well-established molecular biology and genetic techniques (targeted gene disruptions and/or site specific mutagenesis, suppressor genetics, 2-hybrid systems) • advanced "proteome" (all proteins) research • resistant to solvents (e.g. DMSO)	• extraordinary variety • therapeutic applications, oral administration • large and unexploited natural reservoir of unknown substances or known substances without target • toxicity (the lower the concentration at which a compound acts, the more likely that it will exhibit specificity and as a consequence, the less likely that it will have undesired side effects) • "molecular design" (synthesis and scale-up); analogues (structure–activity relationship) • low molecular weight (in favour of maintaining a therapeutically significant concentration of the drug in the vicinity of its target for the desired period of time) • no immune response (no attenuation of the therapeutic benefit and no toxicity resulting from activation of immune-system cascades) • powerful analytical methods (MS, NMR, HPLC, etc.)

O. Groudinsky and A. Baudin for their interest in this work and discussions. We thank Drs F. M. Klis and J. Rytka for suggestions concerning growth conditions and M.-L. Bourbon, F. Casalinho, P. Kerboriou, M. C. Lucinus for their availability and help in preparation of various standard media.

References

Aguilera, A. (1994). Formamide sensitivity: a novel conditional phenotype in yeast. *Genetics* **136**, 87–91.

Anderson, J. A., Huprikar, S. S., Kochian, L. V., Lucas, W. J. and Gaber, R. F. (1992). Functional expression of a probable *Arabidopsis thaliana* potassium channel in *Saccharomyces cerevisiae*. *Proc. Natl. Acad. Sci. USA* **89**, 3736–3740.

Ballou, L., Hitzeman, R. A., Lewis, M. S. and Ballou, C. E. (1991). Vanadate-resistant yeast mutants are defective in protein glycosylation. *Proc. Natl. Acad. Sci. USA* **88**, 3209–3212.

Battaner, E. and Vazquez, D. (1971). Inhibitors of protein synthesis by ribosomes of the 80-S type. *Biochim. Biophys. Acta* **254**, 316–330.

Baudin, A., Ozier-Kalogeropoulos, O., Denouel, A., Lacroute, F. and Cullin, C. (1993). A simple and efficient method for direct gene deletion in *Saccharomyces cerevisiae*. *Nucl. Acids. Res.* **21**, 3329–3330.

Baudin-Baillieu, A., Guillemet, E., Cullin, C. and Lacroute F. (1997). Construction of a yeast strain deleted for the *TRP1* promoter and coding region that enhances the efficiency of the polymerase chain reaction-disruption method. *Yeast* **13**, 353–356.

Beach, D. H., Rodgers, L. and Gould, J. (1985). RAN1+ controls the transition from mitotic division to meiosis in fission yeast. *Curr. Genet.* **10**, 297–311.

Borst-Pauwels, G. W. F. H. (1981). Ion transport in yeast. *Biochim. Biophys. Acta* **650**, 88–127.

Bossemeyer, D., Schlösser, A. and Bakker, E. P. (1989). Specific cesium transport via the *Escherichia coli* Kup (TrkD) K^+ uptake system. *J. Bacteriol.* **171**, 2219–2221.

Bossier, P., Fernandes, L., Rocha, D. and Rodrigues-Pousada, C. (1993). Overexpression of *YAP2*, coding for a new yAP protein, and *YAP1* in *Saccharomyces cerevisiae* alleviates growth inhibition caused by 1,10-phenanthroline. *J. Biol. Chem.* **268**, 23 640–23 645.

Boucherie, H., Dujardin, G., Kermorgant, M., Monribot, C., Slonimski, P. and Perrot, M. (1995). Two-dimensional protein map of *Saccharomyces cerevisiae*: Construction of a gene–protein index. *Yeast* **11**, 601–613.

Broach, J. R. and Thorner, J. (1996). High-throughput screening for drug discovery. *Nature* **384** (Suppl.), 14–16.

Bruno, N. A. and Slate, D. L. (1990). Effect of exposure to calcium entry blockers on doxorubicin accumulation and cytotoxicity in multidrug-resistant cells. *J. Natl. Cancer Inst.* **82**, 419–424.

Burns, N., Grimwade, B., Ross-Macdonald, P. B., Choi, E.-Y., Finberg, K., Roeder, G. S. and Snyder, M. (1994). Large-scale analysis of gene expression, protein localization, and gene disruption in *Saccharomyces cerevisiae*. *Genes Dev.* **8**, 1087–1105.

Codani, J. J., Comet, J. P., Ande, J. C., Glemet, E., Wozmiak, A., Risler, J. L., Henaut, A. and Slonimski, P.P. (1999). In *Methods in Microbiology, vol. 28* (A. Craig and J. D. Hoheisel, eds), pp. 229–244, Academic Press, London, in press.

Conklin, D. S., Kung, C. and Culbertson, M. R. (1993). The *COT2* gene is required for glucose-dependent divalent cation transport in *Saccharomyces cerevisiae*. *Mol. Cell. Biol.* **13**, 2041–2049.

Conklin, D. S., Culbertson, M. R. and Kung, C. (1994). Interactions between gene products involved in divalent cation transport in *Saccharomyces cerevisiae*. *Mol. Gen. Genet.* **244**, 303–311.

Coppée, J.-Y., Rieger, K.-J., Kaniak, A., Di Rago, J.-P., Groudinsky, O. and Slonimski, P. P. (1996). *PetCR46*, a gene which is essential for respiration and integrity of the mitochondrial genome. *Yeast* **12**, 577–582.

Cullin, C. and Minvielle-Sebastia, L. (1994). Multipurpose vectors designed for the fast generation of N- or C-terminal epitope-tagged proteins. *Yeast* **10**, 105–112.

Dancis, A., Yuan, D. S., Haile, D., Askwith, C., Eide, D., Moehle, C., Kaplan, J. and Klausner, R. D. (1994). Molecular characterization of a copper transport protein in *S. cerevisiae*: An unexpected role for copper in iron transport. *Cell* **76**, 393–402.

Dujon, B. (1981). Mitochondrial genetics and functions. In *The Molecular Biology of the Yeast* Saccharomyces (J. N. Strathern, E. W. Jones and J. R. Broach, eds), pp. 505–635. Cold Spring Harbor, New York.

Exinger, F. and Lacroute, F. (1992). 6-Azauracil inhibition of GTP biosynthesis in *Saccharomyces cerevisiae*. *Curr. Genet.* **22**, 9–11.

Farkas, V. (1989). Protein synthesis. In *The Yeasts*, vol. 3 (A. H. Rose and J. S. Harrison, eds), pp. 317–366. Academic Press, London.

Farrell, R. E., Germida, J. J. and Huang, P. M. (1993). Effects of chemical speciation in growth media on the toxicity of mercury-II. *Appl. Environ. Microbiol.* **59**, 1507–1514.

Fernandez, M., Fernandez, E. and Rodicio, R. (1994). *ACR1*, a gene encoding a protein related to mitochondrial carriers, is essential for acetyl-CoA synthetase activity in *Saccharomyces cerevisiae*. *Mol. Gen. Genet.* **242**, 727–735.

Friedman, S. (1982). Bactericidal effect of 5-azacytidine on *Escherichia coli* carrying *Eco*RII restriction-modification enzymes. *J. Bacteriol.* **151**, 262–268.

Gaxiola, R., de Larrinoa, I. F., Villalba, J. M. and Serrano, R. (1992). A novel and conserved salt-induced protein is an important determinant of salt tolerance in yeast. *EMBO J.* **11**, 3157–3164.

Gennis, R. B. (1989). Interactions of small molecules with membranes: partitioning, permeability, and electrical effects. In *Biomembranes* (C. R. Cantor, ed.), pp. 235–269. Springer-Verlag, New York.

Georgatsou, E. and Alexandraki, D. (1994). Two distinctly regulated genes are required for ferric reduction, the first step of iron uptake in *Saccharomyces cerevisiae*. *Mol. Cell. Biol.* **14**, 3065–3073.

Goffeau, A., Barrell, B. G., Bussey, H. *et al.* (1996). Life with 6000 genes. *Science* **274**, 562–567.

Henderson, G. E., Evans, I. H. and Bruce, I. J. (1989). Vanadate inhibition of mitochondrial respiration and H^+ ATPase activity in *Saccharomyces cerevisiae*. *Yeast* **5**, 73–77.

Hogan, J. C. (1996). Directed combinatorial chemistry. *Nature* **384** (Suppl.), 17–19.

Iida, H., Sakaguchi, S., Yagawa, Y. and Anraku, Y. (1990). Cell cycle control by Ca^{2+} in *Saccharomyces cerevisiae*. *J. Biol. Chem.* **265**, 21 216–21 222.

Jackson, C. L. and Képès, F. (1994). *BFR1*, a multicopy suppressor of brefeldin A-induced lethality, is implicated in secretion and nuclear segregation in *Saccharomyces cerevisiae*. *Genetics* **137**, 423–437.

Janis, R. A., Silver, P. J. and Triggle, D. J. (1987). Drug action and cellular calcium regulation. *Adv. Drug Res.* **16**, 309–591.

Kahn, P. (1995). From genome to proteome: looking at a cell's proteins. *Science* **270**, 369–370.

Kosman, D. J. (1994). Transition metal ion uptake in yeasts and filamentous fungi. In *Metal Ions in Fungi* (G. Winkelmann and D. R. Winge, eds), pp. 1–38. Marcel Dekker, New York.

Kuge, S. and Jones, N. (1994). YAP1 dependent activation of *TRX2* is essential for the response of *Saccharomyces cerevisiae* to oxidative stress by hydroperoxides. *EMBO J.* **13**, 655–664.

Lamond, A. I. and Mann, M. (1997). Cell biology and the genome projects – a concerted strategy for characterizing multiprotein complexes by using mass spectrometry. *Trends Cell Biol.* **7**, 139–142.

Li, R. and Murray, A.W. (1991). Feedback control of mitosis in budding yeast. *Cell* **66**, 519–531.

Manfredi, J. J. and Horwitz, S. B. (1984). Taxol: an antimitotic agent with a new mechanism of action. *Pharmac. Ther.* **25**, 83–125.

Meyers, S., Schauer, W., Balzi, E., Wagner, M., Goffeau, A. and Golin, J. (1992). Interaction of the yeast pleiotropic drug resistance genes *PDR1* and *PDR5*. *Curr. Genet.* **21**, 431–436.

Nakai, K. and Kanehisa, M. (1992). A knowledge base for predicting protein localization sites in eukaryotic cells. *Genomics* **14**, 897–911.

Neigeborn, L. and Carlson, M. (1987). Mutations causing constitutive invertase synthesis in yeast: genetic interactions with *snf* mutations. *Genetics* **115**, 247–253.

Oliver, S. G. and Warmington J. R. (1989). Protein synthesis. In *The Yeasts*, vol. 3 (A. H. Rose and J. S. Harrison, eds), pp. 117–160. Academic Press, London.

Oliver, S. G., van der Aart, Q. J., Agostini-Carbone, M. L. *et al.* (1992). The complete DNA sequence of yeast chromosome III. *Nature* **357**, 38–46.

Omura, S. (1981). Lipids. In *Methods in Enzymology*, vol. 72 (J. M. Lowenstein, ed.), pp. 520–532. Academic Press, New York.

Parsons, W. J., Ramkumar, V. and Stiles, G. L. (1988). Isobutyl-methylxanthine stimulates adenylate cyclase by blocking the inhibitory regulatory protein G_i. *Mol. Pharmacol.* **34**, 37–41.

Perkins, J. and Gadd, G. M. (1993). Caesium toxicity, accumulation and intracellular localization in yeasts. *Mycol. Res.* **97**, 717–724.

Petersen, J. G. L., Kielland-Brandt, M. C., Nilsson-Tillgren, T., Bornaes, C. and Holmberg, S. (1988). Molecular genetics of serine and threonine catabolism in *Saccharomyces cerevisiae*. *Genetics* **119**, 527–534.

Ram, A. F. J., Wolters, A., Ten Hoopen, R. and Klis, F. M. (1994). A new approach for isolating cell wall mutants in *Saccharomyces cerevisiae* by screening for hypersensitivity to calcofluor white. *Yeast* **10**, 1019–1030.

Rieger, K.-J., Kaniak, A., Coppée, J.-Y., Aljinovic, G., Bandin-Bailleu, A., Orlowska, G., Gromadka, R., Groudinsky, O., Di Rago, J.-P. and Slonimski, P. P. (1997). Large-scale phenotypic analysis – the pilot project on yeast chromosome III. *Yeast* **13**, 1547–1562.

Romandini, P., Tallandini, L., Beltramini, M., Salvato, B., Manzano, M., de Bertoldi, M. and Rocco, G. P. (1992). Effects of copper and cadmium on growth, superoxide dismutase and catalase activities in different yeast strains. *Comp. Biochem. Physiol.* **103C**, 255–262.

Rose, M. D., Winston, F. and Hieter, P. (1990). *Methods in Yeast Genetics: A Laboratory Course Manual*. Cold Spring Harbor Laboratory Press, NY.

Ruhland, A. and Brendel, M. (1979). Mutagenesis by cytostatic alkylating agents in yeast strains of differing repair capacities. *Genetics* **92**, 83–97.

Schindler, D. and Davies, J. (1975). Inhibitors of macromolecular synthesis in yeast. *Meth. Cell Biol.* **12**, 17–38.

Sipos, G., Puoti, A. and Conzelmann, A. (1994). Glycosylphosphatidylinositol membrane anchors in *Saccharomyces cerevisiae*: absence of ceramides from complete precursor glycolipids. *EMBO J.* **13**, 2789–2796.

Slater, F. C. (1973). The mechanism of action of the respiratory inhibitor, antimycin. *Biochim. Biophys. Acta* **301**, 129–154.

Slonimski, P. P. and Brouillet, S. (1993). A data-base of chromosome III of *Saccharomyces cerevisiae*. *Yeast* **9**, 941–1029.

Smith, V., Chou, K. N., Lashkari, D., Botstein, D. and Brown, P. O. (1996). Functional analysis of the genes of yeast chromosome V by genetic footprinting. *Science* **274**, 2069–2074.

Thomas, B. J. and Rothstein, R. (1989). Elevated recombination rates in transcriptionally active DNA. *Cell* **56**, 619–630.

Toda, T., Shimanuki, M. and Yanagida, M. (1991). Fission yeast genes that confer resistance to staurosporine encode an AP-1-like transcription factor and a protein kinase related to the mammalian ERK1/MAP2 and budding yeast *FUS3* and *KSS1* kinases. *Genes Dev.* **5**, 60–73.

Torres, A., Rossignol, M. and Beisson, J. (1991). Nocodazole-resistant mutants in *Paramecium*. *J. Protozool.* **38**, 295–304.

Treinin, M. and Simchen, G. (1993). Mitochondrial activity is required for the expression of *IME1*, a regulator of meiosis in yeast. *Curr. Genet.* **23**, 223–227.

Tuite, M. F. (1989). Protein synthesis. In *The Yeasts*, vol. 3 (A. H. Rose and J. S. Harrison, eds), pp. 161–204. Academic Press, London.

Van der Leij, I., Van den Berg, M., Boot, R., Franse, M., Distel, B. and Tabak, H. F. (1992). Isolation of peroxisome assembly mutants from *Saccharomyces cerevisiae* with different morphologies using a novel positive selection procedure. *J. Cell Biol.* **119**, 153–162.

Velculescu, V. E., Zhang, L., Zhou, W., Vogelstein, J., Basrai, M. A., Bassett Jr., D. E., Hieter, P., Vogelstein, B. and Kinzler, K. W. (1997). Characterization of the yeast transcriptome. *Cell* **88**, 243–251.

Verdine, G. L. (1996). The combinatorial chemistry of nature. *Nature* **384** (Suppl.), 11–13.

Xu, H. and Shields, D. (1993). Prohormone processing in the trans-golgi network: endoproteolytic cleavage of prosomatostatin and formation of nascent secretory vesicles in permeabilized cells. *J. Cell Biol.* **122**, 1169–1184.

Yoshida, S., Ikeda, E., Uno, I. and Mitsuzawa, H. (1992). Characterization of a staurosporine- and temperature-sensitive mutant, *stt1*, of *Saccharomyces cerevisiae*: *STT1* is allelic to *PKC1*. *Mol. Gen. Genet.* **231**, 337–344

10 Automatic Analysis of Large-scale Pairwise Alignments of Protein Sequences

J. J. Codani[1], J. P. Comet[1], J. C. Aude[1], E. Glémet[1], A. Wozniak[1], J. L. Risler[2], A. Hénaut[2] and P. P. Slonimski[2]

[1] *INRIA Rocquencourt, Le-Chesnay Cedex, France;* [2] *Centre de Génétique Moléculaire du CNRS, Gif-sur-Yvette, France*

CONTENTS

I. INTRODUCTION

The aim of this chapter is to describe a set of automatic tools and methods for the analysis of large sets of protein sequences. The amount of data discovered by the genomic analyses are already quite considerable and are increasing very rapidly. One of the main questions, which has been discussed in hundreds of reports and review articles, concerns the estimation of the similarity between protein sequences and their classification into groups of similarity. The approaches presented here are in many ways different from those used most frequently. The significance of a similarity is estimated by a Monte-Carlo simulation and the allocation into similarity groups is performed by a continuous probability threshold scanning. Furthermore, the individual similarity groups are analysed by a hierarchical clustering method, where any object can be linked to two other objects, which has not been used until now in proteinology.

All protein sequences coded by five completely sequenced microbial genomes have been aligned pairwise. Similar sequences have been grouped into clusters of paralogs (coded by the same genome) and orthologs (coded by different genomes). As a result, intra- and inter-genome families of

METHODS IN MICROBIOLOGY, VOLUME 28
0580-9517 $30.00

proteins have been constructed. Unexpected and challenging results have been obtained in terms of biological and evolutionary implications which are reported elsewhere (Slonimski *et al.*, 1998). Here we summarise the bioinformatics part of this study. A more in-depth description of the methods can be found in Comet *et al.* (1998) and Aude *et al.* (1998), and a more detailed description of LASSAP software used in Glémet and Codani (1997).

In order to classify efficiently tens of thousands of proteins (leading to hundreds of millions of pairwise alignments) one needs powerful computation tools and a robust probability model to estimate the significance of a pairwise alignment. From the probabilities we can induce a similarity/dissimilarity index between two sequences. We can therefore build clusters of related sequences, and apply classification algorithms on each of them.

This chapter is divided into four further sections and a conclusion.

Section II details LASSAP, a new sequence comparison package designed to overcome some limitations of current implementations of sequence comparison programs, and designed to fit the needs of large-scale analysis.

Section III details the Z-value method and focuses on a statistical analysis of the distribution of Z-values. For real proteins, we observe an over-representation of high Z-values, in comparison with sequences of the same length and amino acid composition generated by random shuffling of real sequences (which we shall call henceforth "quasi-real" sequences). Thus, if the significance of an alignment score is based on the theoretical Extreme Value Distribution (which fits well the "quasi-real" sequences), then the significance of high Z-values will be overestimated. We determine first a cut-off value which separates these overestimated Z-values from those which follow the Gumbel distribution. We then show that the interesting zone of distribution of Z-values can be approximated by a Gumbel distribution with different parameters or by a Pareto law.

Section IV details some of the parameters and data used to analyse five complete microbial genomes: *Saccharomyces cerevisiae, Haemophilus influenzae, Methanococcus jannaschii, Synechocystis, Escherichia coli.*

Section V deals with the pyramidal classification method used to analyse each cluster individually.

◆◆◆◆◆◆ II. LARGE-SCALE SEQUENCE COMPARISON PACKAGE (LASSAP)

Current implementations of sequence comparison programs have been designed in a "single algorithm, single sequence query, single database, one shot" spirit. Therefore, taken as a whole, these implementations (although powerful as individual queries) suffer from several weaknesses. Indeed, there is no easy way: (i) to deal with multiplicity of data formats; (ii) to compare results from different algorithms; (iii) to compute and analyse alignments at a database comparison level; (iv) to post-process results. In order to overcome these limitations, INRIA has designed and implemented a software package called LASSAP.

LASSAP presents a new approach to sequence comparison. It consists of a *kernel* and a *set of algorithms*. The kernel provides a simple way to add any pairwise-based algorithm through an API (Application Programming Interface). Thus, LASSAP is a framework allowing the integration of new algorithms. The kernel also provides numerous services shared by all algorithms. As a result, LASSAP is an integrated software for end-users. LASSAP currently implements all major sequence comparison algorithms and string matching and pattern matching algorithms. LASSAP implements new algorithms *ex-nihilo* or by the combination of existing ones. As an example, Z-value computation has been integrated into LASSAP in this way.

A complete description of LASSAP can be found in Glémet and Codani (1997).

A. LASSAP Foundations

A study of the overall process of sequence comparison shows that whatever algorithm is used, the process can be split into four independent treatments:

1. *Input management*: this includes command line parsing, scoring matrices and databank handling which can also be decomposed into three states: loading data, selecting subsets, and translating in frames (in the case of nucleic acid sequences).
2. *Computation*: as a first approximation, a computation between two sequences is formed of a pairwise sequence comparison algorithm, which includes the initialisation of parameters of the algorithm, the algorithm itself, and the appropriate post-treatments.
3. *Control flow*: it controls the global computation (sequence against databank, databank against databank, . . .) by looping on all pairwise comparisons induced by data.
4. *Output management*: this involves the filtering and storing of results. One has to note that every kind of algorithm computing alignments produce results, which can be stored using the same data structure.

This is the reason why LASSAP has been designed in a modular way as illustrated by Figure 1.

An algorithm in LASSAP interacts with the kernel (modules 1, 3 and 4), and any enhancements of these modules benefit the algorithm. The following subsections detail the services provided by the kernel to any pairwise algorithm.

B. Complex Queries

LASSAP allows "complex queries" in the following sense: a databank (or a query) can be a whole databank or a subset of a databank obtained through a selection mechanism. Frame (or phase) translations can be applied to both of them. It is possible to compare a databank against itself or against another databank. This integrated feature avoids having to launch numerous "sequence against databank" programs, and also avoids having to deal with numerous result files. Above all, this feature, combined with structured results, is the best way to perform complex post-analysis. Moreover, it allows an efficient parallel implementation.

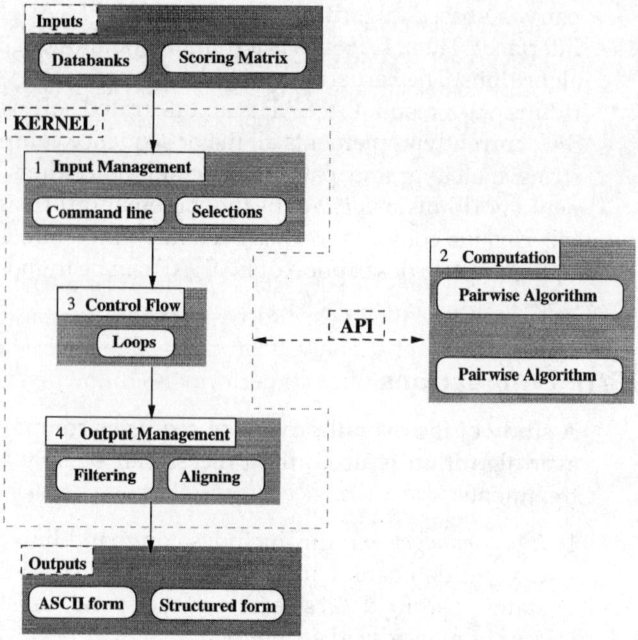

Figure 1. The modular architecture of LASSAP. Each module is in charge of specialised treatments (modules 1 to 4).

It is also very useful to select a subset of a databank and compute the result on the fly. Selections in LASSAP are regular expressions operating on headers and/or sequences and lengths. External query systems, such as SRS (Etzold and Argos, 1993), can also be called by LASSAP. Lastly, a LASSAP databank (or selection) can be translated on the fly into DNA reverse complementary or into frames. The genetic code can be specified in the command line.

C. Performance Issues

As already stated, performance improvements are necessary for rigorous sequence analysis. There are two ways to reduce the computation time: (i) paralleling the algorithm itself; (ii) paralleling the external loops, by taking into account the independence of comparisons. The first solution is well suited for regular algorithms such as dynamic programming. The second solution can be implemented by software on parallel architectures (parallel computers, workstation networks, ...); in this case, each processor in the parallel machine computes a part of the iteration space (the set of all pairs of sequences to be compared). This is achieved by the Control-flow module of LASSAP which handles both cases:

- *Parallel architectures.* Whatever the algorithm, the LASSAP module in charge of the control flow provides automatic spreading of the computation on shared memory and message passing architectures.
- *The Algorithm itself.* An optimised implementation of the Smith–Waterman algorithm has been devised (Wozniak, 1997) using the visual instruction set of

the Sun UltraSparc processor. Performance reaches 35 million matrix cell updates per second (MCUS) on a single 300 MHz UltraSparc processor.

By combining the two points described above, performance reaches hundreds of MCUS on multiprocessor servers.

D. Structured Results

Alignments in LASSAP can be displayed in ASCII form, in a format close to the usual ones (blast, fasta, ...). They can also be stored as a structured binary file and then post-processed. The advantage of structured results is that multiple post-analysis of results can be carried out without a new run. For example, one can perform the following:

- Various simple post-analyses on the properties of the alignments, such as sorting by scores, by probabilities, by lengths of alignments, etc. Moreover, one can extract alignments by a selection combining criterion on alignment and sequence properties.
- Complex post-analysis such as the building of clusters of related sequences. This will be detailed in Sections III and IV.
- A multiple alignment with a first pass based on all pairwise comparisons (e.g. clustal, pileup).
- Databank redundancy. LASSAP is used in this way by the EBI to reduce SWISSPROT/TREMBL database redundancy (Apweiler *et al.*, 1997).

E. Implemented Algorithms

From a programmer's point of view, module 2 (Figure 1) is user programmable through an API. The way an algorithm is plugged into LASSAP is described in detail in Glémet and Codani (1997).

LASSAP currently implements all major sequence comparison algorithms: Blast (Altschul *et al.*, 1990), Fasta (Pearson and Lipman, 1988), Dynamic programming with global and local similarity searches (Needleman and Wunsch, 1970; Smith and Waterman, 1981), as well as k-best alignments (Miller and Huang, 1991). Special attention has been given to Z-value implementation and its associated probability (see Section III). Other kinds of useful algorithms for string matching and pattern matching are also implemented. For example, this allows: (i) PROSITE pattern searching on proteins and/or translated DNA; (ii) subfragment searching, with a given percentage of errors which can be insertions, deletions or substitutions.

This list is not exhaustive. Other algorithms, which combine the algorithms above, are implemented. For instance, LASSAP implements an algorithm which computes Smith–Waterman alignments depending on Blast results.

F. Using LASSAP

LASSAP is an integrated software and is not a combination of shell scripts. It allows one to choose an algorithm as a parameter of the sequence comparison process. The chosen method is a parameter of the command line (-M flag).

For example, the following command line launches the main program `lspcalc`, and computes Z-values (`-M ZValue`), with BLOSUM62 matrix, between two databanks: the first one is composed of yeast sequences from SWISSPROT (_YEAST in SWISSPROT IDs) whose lengths are greater than 500 amino acids (`H.ID _YEAST` and `L > 500`); this is a LASSAP selection); the second one (the query) is the prokaryota section of EBML databank on which phase translation is applied on the three positive frames (`-f top`). A cut-off score is specified (`-scut 6`). Results are stored under the binary file res (`-o res`). The computation runs on eight processors (`-P 8`).

```
% lspcalc
   -M ZValue
   -mp BLOSUM62
   -db swissprot {H.ID _YEAST and L > 500}
   -db -f top /db/embl/pro
   -scut 6
   -o res
   -P 8
```

Once done, results can be post-analysed in various ways using `lspread` program. For example, the following command line:

```
% lspread res ((Z > 8) or (PI > 25)) and (HQuery "heat shock")
```

retrieves alignments whose Z-values are greater than 8 or percentage of identity is greater than 25 and which implies heat shock genes.

This example shows some capabilities of LASSAP, which can imply a quite complicated command line. The VLASSAP tool, is a Java front-end for LASSAP which allows a user-friendly interaction and displays results in a graphical mode.

◆◆◆◆◆◆ III. Z-VALUE

The first adaptation of dynamic programming for sequence alignment was due to Needleman and Wunsch (1970), and subsequent improvements and extensions were made by Smith and Waterman (1981), Waterman and Eggert (1987) and Miller and Huang (1991). Any alignment of two protein sequences by these algorithms results in a so-called optimal alignment score. Nevertheless, the optimality of the score does not ascertain that the two sequences are indeed related. Numerous reports focus on the expression of a probability that the score could be obtained by chance. For non-gapped alignments, such as Blast, a theoretical model exists. It does not apply for gapped alignments. One can refer to Mott (1992), which describes a method for estimating the distribution of scores obtained from a databank search using the Smith and Waterman algorithm taking into account length and composition of sequences in the distribution function. An interesting approach by Waterman and Vingron (1994) gives an estimation of the significance of the score of a gapped alignment. The authors use the Poisson clumping heuristic to describe the behaviour of scores: as a result, the probability for a score to be lower than or equal to t is approximately exp $(-\gamma mnp^t)$, where m, n are the sequence length, and γ and p are parameters estimated on data.

A complementary approach is to use the Z-value. The Z-value relies on a Monte-Carlo evaluation of the significance of the Smith–Waterman score (Landes *et al.*, 1992; Lipman *et al.*, 1984; Slonimski and Brouillet, 1993). The method consists of comparing one of the two sequences with some randomly shuffled versions of the second one (Lipman and Pearson, 1985). The shuffled sequences share exactly with the initial second sequence the same amino acid composition and length. This simulation takes into account the bias due to the amino acid composition, and partly to the length. This method is used in the RDF2 package (Karlin and Altschul, 1990) and other programs like Bestfit (Devereux, 1989).

Given two sequences A and B, and the Smith–Waterman score $S(A, B)$, the method aims to perform N comparisons between the first sequence A and N shuffled sequences from B, which yield the empirical mean score $\tilde{m}$ and the empirical standard deviation $\tilde{\sigma}$. The Z-value Z is then defined as:

$$Z(A,B) = \frac{S(A,B) - \tilde{m}}{\tilde{\sigma}}$$

For this shuffling process, the exact number N of *shuffled sequences* is so large that the computation of the mean and the standard deviation is not practically feasible over all the possible shuffled sequences. Moreover, the Z-value can depend on the choice of the shuffled sequence (A or B). An in-depth study (Comet *et al.*, 1998) led us to take $N = 100$ and $Z_{\text{value}} = min\,(Z(A, B), Z(B, A))$. Using Z-values rather than Smith–Waterman scores obviously leads to different results. Figure 2 reports the quantitative differences observed between scores and Z-values at a whole genome comparison level. It highlights the non correlation between scores and Z-values in the "twilight zone", i.e. the range of scores in between high scores and low scores. For very high scores, which represent a very small fraction of all possible alignments (less than 0.001), a reasonably good correlation with corresponding Z-values is observed; therefore, the sequences are obviously related. However, for scores that occur with frequencies higher than 0.001, no correlation is found. The significance of a pairwise alignment method relies precisely on its ability to give a reliable decision concerning the similarity between sequences in the twilight zone. It is important to stress that, although the "twilight zone" represents a small fraction of all the pairwise alignments (of the order of 2%), the fraction of proteins involved in it may be quite large (of the order of 50% of a single genome).

A. Statistical Analysis of the Distribution of Z-values

The aim of this study is to find a law of probability that the experimental Z-values will follow. Indeed, from a probability, we can induce a similiarity/dissimilarity index between two sequences. We can therefore build clusters of related sequences, and apply classification algorithms on each of them.

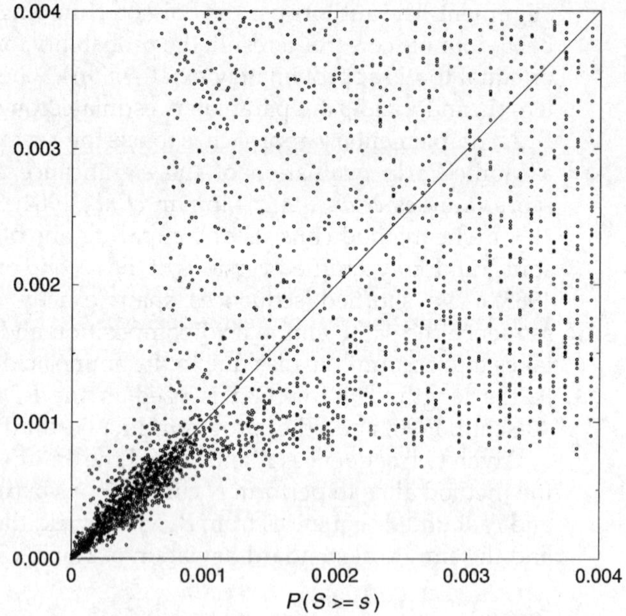

Figure 2. Non-correlation between frequency distribution of SW-scores and corresponding Z-values in the "twilight zone". All alignments for proteins coded by the *Haemophilus influenzae* genome have been computed. Each alignment has a Smith–Waterman score and a Z-value, with associated probabilities. For a genome of size *N*, *C* alignments are computed ($C = N * (N - 1)/2$). The *score* probability, which is the expectancy to have a score *S* greater than or equal to *s*, is defined as: $P(S \geq s)$ = (*number of observed scores greater or equal than s*)/C. *Z-value* probabilities are defined in the same way. Any alignment is then defined by two co-ordinates (these two probabilities). This figure reports the set of co-ordinates of alignments whose probabilities are bound in the frame (0, 0.004). If Z-value and scores were equivalent, all points should be placed near the first diagonal. This is true for very low probabilities (scores and Z-values are high), but a dispersion begins in the neighbourhood of point (0.0008, 0.0008) – a Z-value probability of 0.0008 corresponds to a Z-value of 9. This figure highlights the set of alignments with high SW-score, low-Z-value and vice versa.

In a more detailed study (Comet *et al.*, 1998), various parameters of the Z-value have been analysed, more precisely the Gumbel distribution (Gumbel, 1958). This has to be correlated with studies of Karlin, Altschul and co-workers (Altschul *et al.*, 1990; Karlin and Altschul, 1990; Karlin *et al.*, 1990), which have shown that distribution of Blast scores for sequences of independent identically distributed letters follows the Extreme Value Distribution (EVD, type *I*). Briefly, for two random sequences $A = a_1a_2\ldots a_n$ and $B = b_1b_2\ldots b_m$, given the distribution of individual residues, and given a scoring matrix, the probability of finding a segment pair with a score greater than or equal to *s* is:

$$P(X \geq s) = 1 - \exp\left(-K.m.n.\exp^{-\lambda.s}\right)$$

where λ and *K* may be calculated from the substitution matrix and sequences composition.

For the estimation of the law of Z-values, we want to find two parameters, the characteristic value θ and the decay value ξ, such as:

$$P(Z \geq z) = 1 - \exp\left(-\exp^{-(z-\xi)/\theta}\right)$$

where z is the observed Z-value.

The two parameters θ and ξ, have been estimated with the Maximum Likelihood Estimators (Johnson and Kotz, 1970), using large datasets of real sequences (R) and random ones (i.e. shuffled sequences from R). For these parameters, it has been checked that:

- In the case of "quasi-real" sequences, the EVD model is a good estimation of the observed distribution, with parameters $\varepsilon = -0.7$ and $\theta = 0.81$.
- In contrast, for real protein sequences, the EVD model fits quite well the oberved distribution for Z-values lower than 8, with parameters similar to those calculated for "quasi-real" sequences, but is not satisfactory for high Z-values. There are about 1 out of 1000 over-represented Z-values. This over-representation of high Z-values can lead to wrong values of their significance (i.e. the probability $P(Z \geq z_0)$ that one could obtain a Z-value greater or equal to a value z_0). This is illustrated by Figure 3.

Figure 3 shows that real sequences *are not* random sequences. The curves diverge beyond a certain value of Z-value c. That means that Z-values above c are not obtained by chance. This value, c, will be called the *cut-off value*. Figure 4 shows that we can adopt the value 8.0 as a conservative estimate of the cut-off.

This purely formal conclusion is obvious in biological terms. Real protein sequences result from evolution where gene duplications, mutations, fusions and recombinations take place continuously as major forces

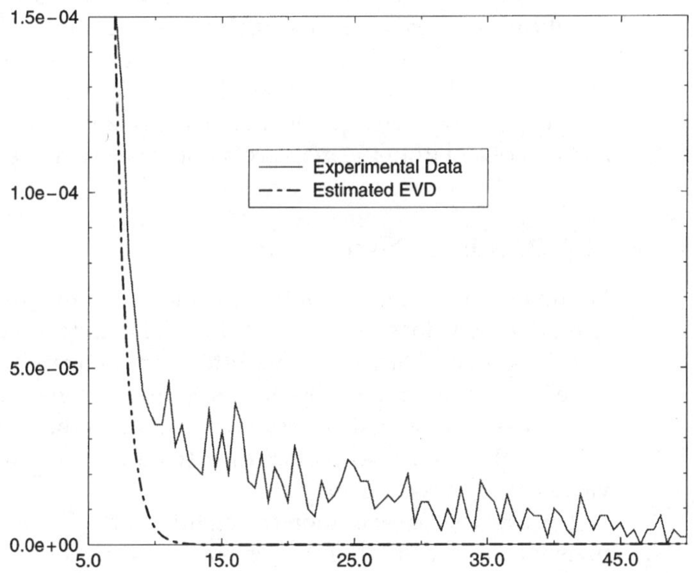

Figure 3. Z-values frequency distribution. Solid line shows observed frequencies of Z-values obtained on a large dataset of yeast genome. Dashed line shows the best approximated Extreme Value Distribution (EVD). For high Z-values, the EVD overestimates the significance of Z-values while it fits quite well the low Z-values.

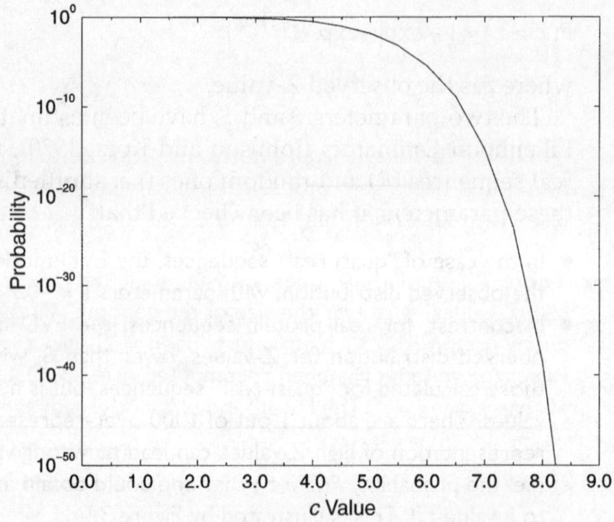

Figure 4. Cut-off value. Estimation of cut-off value for splitting the EVD like Z-values from high Z-values. Let $X \sim B(N, p_c)$ be a binomial variable, where N is the number of observed Z-values and p_c the probability that the EVD variable Z is greater or equal to c. X is the expected number of Z-values greater or equal to c. This figure shows the variation of the probability $P(X > N_{obs}^c)$, where N_{obs}^c is the observed number of Z-values greater than c. The decrease of the probability shows that the observed distribution of real protein sequences diverges from the EVD between 6.0 and 7.0 and becomes practically zero at 8.0. This study has been carried out for both *Haemophilus* and *Methanococcus* genomes and the results are basically the same.

conserving sequence similarities and generating sequence diversities. It should be kept in mind that real protein sequences, those that do exist actually, as well as those that did exist during life's history, represent an infinitely small fraction of all possible random permutations of an average length of 300 with 20 different amino acids (20^{300}). The real protein space is a microcosm within the macrocosm of "quasi-real" sequence space.

B. Law of High Z-values Distribution

We now estimate the law of the Z-values distribution for Z-values greater than 8 (for Z-values lower than 8, the EVD model is kept).

Let us recall that we are interested here in alignments in the "twilight zone" and not in the alignments showing very high Z-value where sequences are obviously very similar (*e.g.* more than 80% of identities over their whole length). To explore this "twilight zone", we considered the Z-values in the range [8,50].

The observed distribution can be fitted with Gumbel law, but the parameters ξ (mean –125) and θ (mean 19.3) are completely different from those of the distribution of Z-values lower than 8 (see supra). In addition, we used linear regression techniques for fitting the distribution curve in the range [8,50]. In that case, the retained model is the Pareto distribution [Zajdenweber, 1996].

The density function of the Pareto distribution is:

$$f(z) = \begin{cases} A.z^{-(1+\alpha)} & \text{if } z \geq 8.0 \\ 0 & \text{otherwise} \end{cases}$$

with $\alpha \geq 0$

The coefficient A is just a normalisation coefficient and is not informative. α is called the Pareto index. Table 1 displays the estimated parameters, for five complete microbial genomes, as well as for all the genomes taken

Table 1. The Pareto index showing that the Pareto law is a good model for high Z-values, whatever the size of the genome. All the indices have been computed using PAM250 matrix (gap open = 5, gap extend = 0.3). *Haemophilus influenzae* genome has been recomputed using BLOSUM62 matrix (gap open = 10, gap extend = 1). The Pareto index is not greatly different.

	Number of pairwise comparisons	Pareto index* α
YR	499 500	1.20
Saccharomyces cerevisiae	18 522 741	0.90
Escherichia coli	9 182 755	1.26
Haemophilus influenzae	1 410 360	1.63
Haemophilus influenzae (BLOSUM62)	1 410 360	1.26
Methanococcus jannaschii	1 504 245	1.16
Synechocystis	5 016 528	1.05
all vs. all	143 744 490	1.16

* α mean = 1.21; standard deviation of the mean = 0.22.

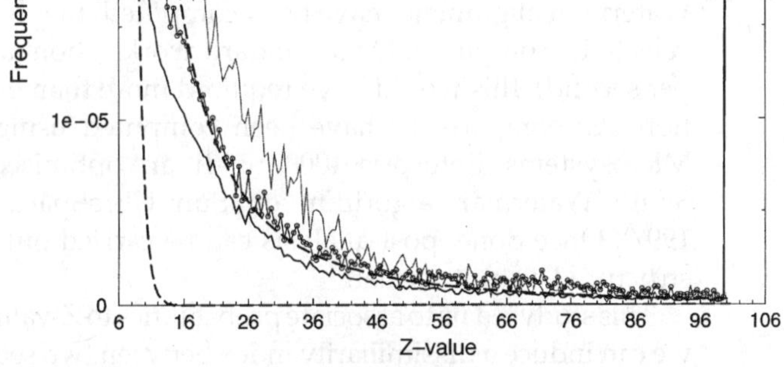

Figure 5. Density of Z-values. For all complete genomes, Z-value density has a non-negligible tail, which differs from the Gumbel distribution valid for Z-values lower than 8 (see Fig. 4). The observed distributions of two genomes (*Escherichia coli* and *S. cerevisiae*) are shown, as well as the observed distribution of five genomes taken all together (All vs. All curve). These distributions are similar and can be fitted by a Pareto law. The Pareto index α is taken as the mean of the estimates for the five genomes (see Table 1).

together. One can observe that for both models, the estimated parameters are independent of the genome size and of the similiarity matrix used in the alignments. In Figure 5 are displayed the experimental distribution of the Z-values together with the Pareto curve.

Moreover, additional tests have been performed on *Haemophilus influenzae* genome using BLOSUM62 scoring matrix. They led to the same conclusion: Z-value distribution using BLOSUM62 fits the Pareto distribution with a Pareto index not greatly different from those computed with PAM 250 matrix.

◆◆◆◆◆◆ IV. APPLICATION: MICROBIAL GENOMES

Some of the methods described above have been used first of all to analyse the complete yeast genome (6087 Open Reading Frames potentially coding for protein sequences, 2878174aa; Web site http://speedy.mips.biochem.mpg.de/mips/yeast) and extended later to the study of four other complete microbial genomes: *Haemophilus influenzae, Methanococcus jannaschii, Synechocystis, Escherichia coli* (see http://www.mcs.anl.gov/home/gaasterl/genomes.html Web site).

Throughout this study we consistently used the same scoring matrix for the Smith and Waterman algorithm, that is, the Dayhoff PAM250 matrix divided by 3 (Risler *et al.*, 1988; Schwartz and Dayhoff, 1979) with gap penalties as follows: gap open = 5 and gap extend = 0.3.

The Smith and Waterman scores have been computed for all possible pairwise alignments of sequences. The work presented here led us to consider the alignments whose Z-values are greater than 8.0. They have been further analysed for the similarities between sequences. Comet *et al*. (1998) let us conclude that, for Smith and Waterman scores lower than 22, the Z-values greater than 8.0 are quasi non-existent. Therefore, the cut-off value of 22 for Smith and Waterman scores has been used to compute Z-values.

In addition to the tests, for the ensemble of the five microbial genomes (16 956 sequences, 6 274 509 amino acids) about 300 million Smith/Waterman alignments have been computed, that is about 30×10^{12} matrix cells to be computed. On a standard workstation (at 10×10^{6} matrix cells per second), this would have required more than one month of computation. All comparisons have been computed using LASSAP on a SUN Microsystems Enterprise4000, with an optimised implementation of Smith/Waterman algorithm on Sun UltraSparc processor (Wozniak, 1997). Once done, post-analysis can be carried out easily using LASSAP structured output format.

This study led us to associate probabilities to Z-values. From a probability, we can induce a dissimiliarity index between two sequences. We can therefore build clusters of related sequences, and apply classification algorithms on each of them. We therefore performed clustering with different probability thresholds, that is, the sequences were grouped in "connective clusters" such that in any given connective cluster, any sequence shares a Z-value greater than a given threshold (or shares a Pareto probability lower than a given threshold) with at least another sequence of the same cluster.

By considering each genome individually, or the five genomes taken all together, this procedure led to thousands of clusters which can be considered as families of protein sequences. Contrary to the usual approach, where a single and arbitrary cut-off value is used to construct the single link connective clusters, we have introduced the "probability-threshold-scanning" approach. The 300 million pairwise alignments are scanned and the connective clusters of similar proteins are constructed for every Z-value or probability threshold. In this manner we construct not just one set of connective clusters linked by a single similarity threshold, but a spectrum of sets by increasing step by step the similarity threshold.

Section V describes the pyramidal classification method used to analyse the resulting clusters.

◆◆◆◆◆◆ V. PYRAMIDAL CLASSIFICATION OF CLUSTERS

Any given connective cluster will contain those proteins which display sequence relationships either by vertical descent from a common ancestor (orthologs in different species and paralogs in the same species) or by horizontal transfer. In addition, some connective clusters will contain sequences that share one or several domains with another multi-domain protein. Once the sequences have been clustered, it is generally convenient to perform a classification of the different members in each cluster in order to obtain an immediate visualisation of the different relationships between the sequences. One often resorts to hierarchical clustering methods such as UPGMA (Sneath and Sokal, 1973) or neighbour-joining (Saitou and Nei, 1987). Nevertheless, when a classification performed by, say, UPGMA or neighbour-joining results in the delineation of several subclasses, then it is difficult, if not impossible, to know which sequences are responsible for the links between the subclasses. This difficulty is particularly striking in the case of multi-domain proteins.

As could be expected, the origin of the problem lies in the classification algorithm itself. In the classical hierarchical clustering methods, any object can be linked to *one*, and only one, other object. When two objects have been agglomerated because they are the closest in the distance matrix, they are eliminated from the matrix and replaced by one single node whose distances to the remaining objects are the mean (or max or min) distances of the two original objects to the others. This algorithm presents a drawback, that is, it does not take into account the fact that is is often reasonable to consider that a given object should be linked to *two* other different objects. This is clearly the case with multi-domain proteins.

Some time ago, Bertrand and Diday (1985) developed a new hierarchical clustering method that they called pyramidal clustering. In their algorithm, any object can be linked to *two* other objects. During the construction of the classification tree, two objects that have just been agglomerated are not eliminated from the distance matrix. Instead, their cluster is *added* to the matrix. A detailed description of the method, and its application to protein sequence classification can be found in Aude *et al.* (1998).

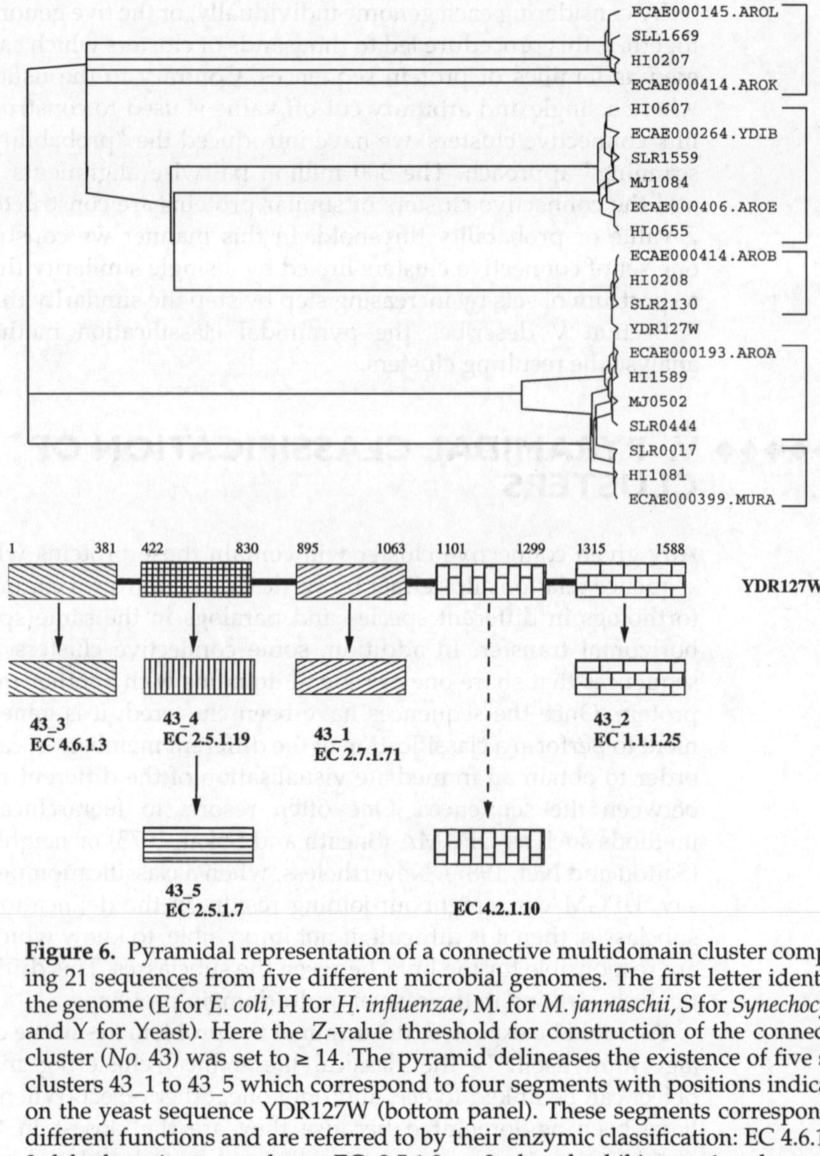

Figure 6. Pyramidal representation of a connective multidomain cluster comprising 21 sequences from five different microbial genomes. The first letter identifies the genome (E for *E. coli*, H for *H. influenzae*, M. for *M. jannaschii*, S for *Synechocystis* and Y for Yeast). Here the Z-value threshold for construction of the connective cluster (*No.* 43) was set to ≥ 14. The pyramid delineases the existence of five subclusters 43_1 to 43_5 which correspond to four segments with positions indicated on the yeast sequence YDR127W (bottom panel). These segments correspond to different functions and are referred to by their enzymic classification: EC 4.6.1.3 = 3-dehydroquinate synthase, EC 2.5.1.9 = 3-phosphoshikimate 1-carboxyvinyl transferase, EC 2.5.1.7 = UDP-N-Acetylglucosamine 1-carboxyvinyl transferase, EC 2.7.1.71 = shikimate kinase, and EC 1.1.1.25 = shikimate 5-dehydrogenase. Note that (*i*) the yeast pentafunctional protein involved in aromatic acid biosynthesis allows to cluster together the different subclusters which otherwise do not display in between them any sequence similiarity; (*ii*) the subcluster 43_4 is highly similar to the yeast sequence (all Z-values are greater than 35), while the subcluster 43_5 is not (all Z-values are smaller than 4). However, the latter subcluster is linked to the yeast sequence *via* the sequences of the subcluster 43_4 to which they display significant similarity (Z-values greater than 14); (*iii*) lowering of the threshold for construction of connective clusters to Z-value = 11, discloses an additional subcluster 43_6 (not shown here), corresponding to the penultimate domain of the yeast protein, comprising two sequences Ecoli.AROD and MJ1454, belonging to the EC 4.2.1.10 class and endowed with 3-dehydroquinate dehydratase activity. Classical methods, such as UPGMA or multiple alignments (Clustal, Phylip), fail in grouping sequences accordingly and lead to erroneous allocations.

We have used this method systematically for all connective clusters. Pyramidal clustering has been performed on the connective clusters with the following definition for the distance $d(i, j)$ between two sequences i and j:

$$d(i,j) = \frac{-\text{Log}(2)}{\text{Log}(Pz(i,j))} \quad \text{if} \quad S(i,j) \geq S_0$$

$$d(i,j) = 1 \text{ otherwise}$$

where $P_z(i, j)$ is the probability associated to the Z-value $Z(i, j)$ for sequences i and j.

One example of a pyramidal classification on a multi-genome connective cluster comprising 21 sequences, is shown in Figure 6.

VI. CONCLUSION

The results presented here show that Z-value computation gives a realistic model to compute probabilities for gapped alignments of protein sequences. It allows the building of reliable clusters of homologous sequences.

The pyramidal classification allows analysis of clusters in a more precise way than commonly used tools, especially in the case of multidomain proteins.

Using sequence comparison tools such as LASSAP, the computation as well as the analysis of large sets of sequence data can be conducted efficiently. Therefore, complete *intra-* and *inter*-genome comparisons and classifications can be carried out as soon as genomes are sequenced and biological implications deduced (Slonimski *et al.*, 1998).

Some results can be accessed at the following address: http://www.gene-it.com.

References

Altschul, S. F., Gish, W., Miller, W., Myers, E. and Lipman, D. (1990). Basic local alignment search tool. *J. Mol. Biol.* **215**, 403–410.

Apweiler, R., Gateau, A., Junker, V., O'Donovan, C., Lang, F., Contrino, S., Martin, M., Mitaritonna, N., Kappus, S. and Bairoch. A. (1997). Protein sequence annotation in the genome era: The annotation concept of swiss-prot + trembl. In: *Fifth International Conference on Intelligent Systems for Molecular Biology* (T. Gaasterland, P. Karp, K. Karplus, C. Ouzounis, C. Sander and A. Valencia eds). ISMB Menlo Park, California. http://www.aaai.org/Press/Proceedings/ISMB/1997/ismb-97.html: AAAI Press.

Aude, J., Diaz-Lazcoz, Y., Codani, J. and Risler, J. (1998). Applications of the pyramidal clustering method to biological object. *Comput.Chem.* Submitted.

Bertrand, P. and Diday, E. (1985). A visual representation of the complexity between an order and a dissimilarity index: the pyramids. *Comput. Stat. Quart.* **2**(1), 31–42.

Comet, J., Aude, J., Glémet, E., Hénaut, A., Risler, J., Slonimski, P. and Codani, J. (1998). An empirical analysis of Z-score statistics generated by large scale pairwise Smith-Waterman alignments of protein sequences. *Comput. Chem.* Submitted.

Devereux, J. (1989). The gcg sequence analysis software package. Package, Version 6.0, Genetics Computer Group Inc., University Research Park, 575 Science Drive, Suite B, Madison, Wisconsin 53711, USA.

Etzold, T. and Argos, P. (1993). SRS – an indexing and retrieval tool for flat file data libraries. *Comp. Appl. BioSci.* **9**, 49–57.
Glémet, E. and Codani, J. (1997). Lassap: a large scale sequence comparison package. *Comp. Appl. BioSci.* **13**(2), 137–143.
Gumbel, E. (1958). *Statistics of Extremes*. Columbia University Press, Columbia.
Johnson, N. L. and Kotz, S. (1970). *Distribution in Statistics: Continuous Univariate Distributions*, vol. 1. The Houghton Mifflin Series in Statistics. The Houghton Mifflin Company, Houghton.
Karlin, S. and Altschul, S. F. (1990). Methods for assessing the statistical significance of molecular sequence features by using general scoring schemes. *Proc. Natl. Acad. Sci.* **87**, 2264–2268.
Karlin, S., Dembo, A. and Kawabata, T. (1990). Statistical composition of high-scoring segments from molecular sequences. *Ann. Stat.* **18**, 571–581.
Landes, C., Hénaut, A. and Risler, J. (1992). A comparison of several similarity indices based on the classification of protein sequences: a multivariate analysis. *Nucl. Acids Res.* **20**(14), 3631–3637.
Lipman, D. and Pearson, W. (1985). Rapid and sensitive protein similarity searches. *Science* **227**, 1435–1441.
Lipman, D., Wilbur, W., Smith, T. and Waterman, M. (1984). On the statistical significance of nucleic acid similarities. *Nucl. Acids Res.* **12**, 215–226.
Miller, W. and Huang, X. (1991). A time-efficient, linear-space local similarity algorithm. *Adv. Appl. math.* **12**, 337–357.
Mott, R. (1992). Maximum-likelihood estimation of the statistical distribution of Smith–Waterman local sequence similarity scores. *Bull. Math. Biol.* **54**(1), 59–75.
Needleman, S. B. and Wunsch, C. D. (1970). A general method applicable to the search for similarities in the amino acid sequence of two proteins. *J. Mol. Biol.* **48**, 443–453.
Pearson, W. R. and Lipman, D. (1988). Improved tools for biological sequence comparison. *Proc. Natl. Acad. Sci. USA* **85**, 2444–2448.
Risler, J.-L., Delorme, M.-O., Delacroix, H. and Hénaut, A. (1988). Amino acid substitution in structurally related proteins. A pattern recognition approach determination of new efficient scoring matrix. *J. Mol. Biol.* **204**, 1019–1029.
Saitou, N. and Nei, M. (1987). The neighbor-joining method: a new method for reconstructing phylogenetic trees. *Mol. Biol. Evol.* **4**, 406–425.
Schwartz, R. and Dayhoff, M. (1979). Matrices for detecting distant relationships. In *Atlas of Protein Sequence and Structure*, vol. 5, suppl. 3, pp. 353–358. National Biomedical Research Foundation, Washington DC.
Slonimski, P. and Brouillet, S. (1993). A database of chromosome III of *Saccharomyces cerevisiae*. *Yeast* **9**, 941–1029.
Slonimski, P., Mossé, M., Golik, P., Hénaut, A., Risler, J., Comet, J., Aude, J., Wozniak, A., Glémet, E. and Codani, J. (1998). The first laws of genomics. *Genom. Comp. Microb.* **3**(1), 46.
Smith, T. and Waterman, M. (1981). Identification of common molecular subsequences. *J. Mol. Biol.* **147**, 195–197.
Sneath, P. and Sokal, R. (1973). *Numerical Taxonomy*. Freeman, San Francisco.
Waterman, M. and Eggert, M. (1987). A new algorithm for best subsequence alignments with application to trna-trna comparisons. *J. Mol. Biol.* **197**, 723–728.
Waterman, M. S. and Vingron, M. (1994). Sequence comparison significance and Poisson approximation. *Stat. Sci.* **9**(3), 367–381.
Wozniak, A. (1997). Using video oriented instructions to speed-up sequence comparison. *Comp. Appl. BioSci.* **13**(2), 145–150.
Zajdenweber, D. (1996). Extreme value in business interruption insurance. *J. Risk Insur.* **63**, 95–110.

11 Towards Automated Prediction of Protein Function from Microbial Genomic Sequences

Michael Y. Galperin[1] and Dmitrij Frishman[2]
[1] *National Center for Biotechnology Information, National Library of Medicine, National Institutes of Health, Bethesda, Maryland, USA;* [2] *Munich Information Center for Protein Sequences/GSF, Martinsried, Germany*

◆◆

CONTENTS

◆◆◆◆◆◆ I. INTRODUCTION

Microbiology, which just a few years ago had to struggle for proper recognition as an independent biological discipline (Woese, 1994), has recently become one the most dynamic branches of biology. This change has been largely due to the availability of complete genome sequences from a number of important and diverse microorganisms. The first two genomes, those of *Haemophilus influenzae* and *Mycoplasma genitalium*, were sequenced at The Institute for Genomic Research (TIGR, Rockville, Maryland, USA) in 1995 (Fleischmann *et al.*, 1995; Fraser *et al.*, 1995). In 1996, four more genomes were completed, including the first representatives of archaea, *Methanococcus jannaschii* (Bult *et al.*, 1996), and eukaryotes, *Saccharomyces cerevisiae* (Goffeau *et al.*, 1996; Mewes *et al.*, 1997a). As it is realistic to expect at least fifty new genomes by the year 2000, the trend still fits the exponential growth pattern. Altogether, about 50 complete genome sequences of bacteria and archaea will probably become available by the year 2000. The lists of sequenced genomes and the projects currently under way are available on the World Wide Web (WWW)*

Automated Prediction of Protein Function

* As most new WWW browsers accept URLs missing the http:// symbol, it is omitted where appropriate.

METHODS IN MICROBIOLOGY, VOLUME 28
0580-9517 $30.00

sites of TIGR (www.tigr.org/tdb/mdb/mdb.html) and Argonne National Laboratory (www.mcs.anl.gov/home/gaasterl/genomes.html).

The success of future attempts to extract the enormous wealth of information contained in complete genomes will largely depend upon our ability to predict protein functions encoded in each genome. Meeting this challenge requires reliable functional predictions for thousands of proteins, a task that is already too complex for manual processing. Hence automation of the sequence analysis process becomes a necessity for any genome sequencing project.

Here we briefly review the database resources and software tools that are currently used in predictions of protein functions and introduce the few integrated software packages that automate database searching and thus greatly simplify the annotator's work. In order to concentrate on the recent developments in the field, we chose to omit the materials that were extensively discussed in several excellent reviews (Altschul *et al.*, 1994; Baxevanis *et al.*, 1997; Bork and Gibson, 1996; Bork and Koonin, 1996; Koonin *et al.*, 1996b) or which will be included in the annual database issue of the *Nucleic Acids Research*, traditionally published each January. Rather, we aimed at providing a working set of current WWW links that would allow the reader to explore the state-of-the-art technologies that are used in genome annotation. This review relies heavily on the availability of the WWW; those without reliable WWW access are encouraged to get the relevant information by E-mail (detailed instructions can be found, e.g. in Baxevanis *et al.*, 1997; Peruski and Peruski, 1997).

◆◆◆◆◆◆ II. MOLECULAR BIOLOGY DATA BANKS

Whenever a piece of DNA is sequenced, the easiest way to find out if it contains a new gene, a variant of an already known gene, or some gene that has been known for years, is to compare it with the enormous body of information deposited in the public data banks. While some important information might be slow to appear in the public domain due to publication constraints, it eventually gets deposited in the public databases and becomes available for everyone's use. A recent encouraging phenomenon has been making the sequence data available to the public even before the formal publication, by releasing them through the WWW. This allows an easy access to the data for anyone with an Internet connection and provides a way to improve sequence annotation through collaboration of various research groups.

A. Nucleic Acid Sequence Databases

The most basic requirement of a nucleic acid sequence database is that it should be comprehensive, up to date and easy to use. The first two objectives are reached through a collaboration of the three major databases: the GenBank, maintained by the National Center for Biotechnology

Information (NCBI) in Bethesda, Maryland, USA; the EMBL Nucleotide Sequence Database at the European Bioinformatics Institute (EBI) in Cambridge, UK; and the DNA Database of Japan (DDBJ) at the National Institute of Genetics in Mishima, Japan. These databases collect new sequence data and exchange updates on a daily basis, so the information kept in each database is basically the same and is arranged using the common principles (listed, e.g. at www.ncbi.nlm.nih.gov/collab). EMBL uses a slightly different format from GenBank and DDBJ, but each nucleotide sequence has the same accession number in all three databases. The information stored in these databases is available to the public by anonymous ftp and through the World Wide Web. In practice, this means that one can connect to any of the three WWW sites, www.ebi.ac.uk/queries/queries.html, www.ddbj.nig.ac.jp/, or www.ncbi.nlm.nih.gov/Entrez, to get the same nucleotide sequence information. All three sites provide the ability to retrieve sequences through text term or sequence similarity searches, although the search engines and interfaces might differ.

It should be noted that databases only serve as repositories of the submitted data; the database staff format the submissions to fit the database standard, but usually do not edit them, unless they contain clear mistakes. On the other hand, the annotations provided by the submitters can later turn out to be imperfect, or even incorrect. From the database curators' point of view, it is the responsibility of the submitters and users to make sure that the annotation is as correct as possible; updates and corrections are always welcome. The record, however, can be changed only by the submitting author(s), who sometimes may be unreachable, or just reluctant to admit past mistakes in sequencing and/or annotation. The few editorial functions that the databases take upon themselves (e.g. removing vector sequences, ensuring correct taxonomic designation of the source organisms, etc.) generate enough controversy to make any further editorial involvement unlikely. In any case, it would be prudent to exercise certain caution before drawing any far-reaching conclusions from a sequence annotation, particularly that of a nucleotide sequence entry in the genome division.

Another important feature of all three databases is that they keep nucleotide sequences in the DNA form, even though they are often derived from mRNA sequences. This means that many genes are represented only by their coding sequences and the introns (if any) are missing. This usually does not pose a problem for bacterial, archaeal or yeast sequences, but has to be considered when using sequences from *Plasmodium falciparum* and higher eukaryotes.

Several specialised databases contain raw sequence data that may be still not deposited in GenBank. While the quality of the data is usually not guaranteed, these data provide a valuable resource for anyone working with these or related organisms. TIGR Microbial Database (www.tigr.org/tdb/mdb/mdb.html), for example, contains not only complete microbial genomes sequenced at TIGR, but also unfinished DNA sequences from *Deinococcus radiodurans*, *Enterococcus faecalis*, *Mycobacterium tuberculosis*, *Neisseria meningitidis*, *Thermotoga maritima*, *Treponema pallidum*, *Vibrio cholerae* and *Plasmodium falciparum*. These data are available for downloading or can be searched for similarity using the NCBI BLAST service (www.ncbi.nlm.nih.gov/BLAST).

The database of the Center for Genome Technology at the University of Oklahoma (www.genome.ou.edu) contains DNA sequences of *Neisseria gonorrhoeae*, *Streptococcus pyogenes*, *Aspergillus nidulans* and *Actinobacillus actinomycetemcomitans*. The genomes of the first two bacteria are almost complete and can also be searched through the NCBI BLAST server. Preliminary sequence data on *Clostridium acetobutylicum*, *Mycobacterium tuberculosis* and *M. leprae* and the finished genome of *Methanobacterium thermoautotrophicum* are available on the Genome Therapeutics Co. WWW site at www.cric.com.

Since the same sequence data may come from several different sources, the NCBI maintains a non-redundant (**nr**) database. This daily updated database is the primary source of data for sequence similarity searching.

B. Protein Sequence Databases

The major sources of protein sequence data are translations of coding sequences from GenBank (GenPept) and EMBL (TrEMBL) and curated protein databases, SWISS-PROT and Protein Identification Resource (PIR).

SWISS-PROT (http://www.expasy.ch/sprot/sprot-top.html), initiated and maintained by A. Bairoch at the Department of Medical Biochemistry, University of Geneva Medical Center in collaboration with the EBI, relies on rigorous sequence analysis of each database entry (Bairoch and Apweiler, 1997). New sequences are included in SWISS-PROT only if there is sufficient evidence that they are correct. In cases of discrepancies between several database entries for the same protein, a combined sequence is included in the database, and the variants are listed in the annotation. SWISS-PROT annotations include descriptions of the function of a protein, its domain structure, post-translational modifications, variants, reactions catalysed by this protein, similarities with other sequences, etc. The enzyme entries contain Enzyme Commission (EC) numbers and are cross-referenced with the ENZYME database (www.expasy.ch/sprot/enzyme.html). The downside of such strict criteria for the database content is the smaller size of SWISS-PROT: it currently contains about 68 000 sequences.

PIR-International is another curated database, maintained by the National Biomedical Research Foundation (http://nbrfa.georgetown.edu/pir/), Munich Information Center for Protein Sequences (www.mips.biochem.mpg.de) and the Japanese International Protein Information Database (George *et al.*, 1997; Mewes *et al.*, 1997b). It contains 98 000 entries that are classified into *ca.* 40 000 protein families and 5000 superfamilies. The MIPS WWW server offers pre-computed multiple sequence alignments at the level of the protein family, protein superfamily, or homology domain. A useful feature of the PIR database is the option to perform complicated queries, such as a search for a protein from a selected species (e.g. *Homo sapiens*) having a certain molecular mass (e.g. from 46 to 48 kDa) or a certain number of residues (e.g. from 246 to 250).

Protein databases at NCBI and EBI include translations of the coding sequences from the respective nucleic acid databases, as well as the data from SWISS-PROT and PIR. These data are merged into a non-redundant (**nr**) database which is used for sequence similarity searches (see Ouellette and Boguski, 1997). The search output always lists the data sources that were used to create each **nr** entry. The total number of protein sequences in **nr** is currently close to 250 000.

C. Motifs, Domains and Families of Homologous Proteins

A protein sequence motif, or pattern, can be broadly defined as a set of conserved amino acid residues that are important for protein function and are located within a certain distance from each other. These motifs often can provide some clues to the functions of otherwise uncharacterised proteins.

The largest and most comprehensive collection of sequence motifs is the PROSITE database (www.expasy.ch/sprot/prosite.html), maintained by A. Bairoch at the University of Geneva Medical Center (Bairoch *et al.*, 1997). This database consists of two files, a textual description of the sequence patterns and protein families characterised by these patterns, and a computer-readable file that allows searching of a given sequence against the patterns in the database. This search can be performed via the WWW interface (www.expasy.ch/sprot/scnpsite.html) or the database (*ca.* 4 Mb) can be downloaded from the PROSITE ftp site and run on a local machine. Instructions for getting the necessary software are posted on the PROSITE WWW site.

Another useful resource for searching protein motifs is the BLOCKS database (www.blocks.fhcrc.org/) developed by Steven Henikoff and co-workers at the Fred Hutchinson Cancer Center in Seattle, WA (Henikoff *et al.*, 1997). Each "block" in this database is a short, ungapped multiple alignment of a conserved region in a family of proteins. These blocks were originally derived from the PROSITE entries, but were later updated using data from many different sources. The BLOCKS server will search a given protein or nucleotide sequence against the blocks in the database; nucleotide sequence will be translated in all six reading frames and each translation will be checked. The BLOCKS database also has an important feature that allows the user to submit a set of sequences, to create a new block and to search this block against the database. This option can be especially useful in cases where a usual database search finds several homologous proteins with no known function. Other databases of sequence motifs, such as PRINTS (www.biochem.ucl.ac.uk/bsm/dbbrowser/PRINTS/PRINTS.html; Attwood *et al.*, 1997) and ProDom (Sonnhammer and Kahn, 1994; Gouzy *et al.*, 1996; http://protein.toulouse.inra.fr), also contain multiple alignments of selected proteins and allow similarity searches against the database. The Pfam database (www.sanger.ac.uk/Software/Pfam/, mirrored at http://pfam.wustl.edu/), was developed by E. Sonnhammer *et al.* (1997) and contains whole protein sequence alignments that were constructed using hidden Markov models (HMMs; Eddy *et al.*, 1995).

A radically different approach to selecting related proteins has been used in the recently created COG database (www.ncbi.nlm.nih.gov/COG) that contains clusters of orthologous groups (COGs) of proteins from each of the completely sequenced genomes (Tatusov *et al.*, 1997). Since orthologs are likely to perform the same function in each organism, identification of an unknown protein as a member of a COG immediately suggests its probable function. As new complete genomes are being constantly added to the COG database, it is likely to become an extremely effective tool for protein function prediction. Similarity searching against the COG database is available at www.ncbi.nlm.nih.gov/COG/cognitor.html.

D. Protein Structure Related Resources

Three-dimensional (3D) protein structures are much harder to determine than primary sequences; they are also much more informative. Knowledge of atomic co-ordinates leads to elucidation of the active site architecture, packing of secondary structural elements, patterns of surface exposure of side-chains and relative positions of individual domains. Structural information is available only for a limited number of proteins, comprising *ca.* 600 distinct protein folds. In completely sequenced genomes only roughly every seventh protein has a known structural counterpart.

The atomic co-ordinates determined by X-ray crystallography and/or NMR spectroscopy are deposited in the Protein Data Bank (www.pdb.bnl.gov) at Brookhaven National Laboratory, which is mirrored at several places, including the Crystallographic Data Centre (http://pdb.ccdc.cam.ac.uk/) in Cambridge, UK, and the Hall Institute of Medical Research in Melbourne, Australia (http://pdb.wehi.edu.au/pdb/).

The Structural Classification Of Proteins database (SCOP, http://scop.mrc-lmb.cam.ac.uk/scop), developed by A. Murzin *et al.* (1995) at the MRC Laboratory of Molecular Biology (Cambridge, UK) and mirrored, e.g. at www.pdb.bnl.gov/scop/, provides a systematic view of the known protein structures. It also offers similarity searching of a given protein sequence against the database, which allows one to determine its nearest relative with known 3D structure. In cases of sufficient sequence similarity such comparison may yield important structural information. The HSSP database (Sander and Schneider, 1991) at http://www.embl-heidelberg.de/srs5 contains multiple sequence alignments of different proteins, at least one of which has known 3D structure: this augments the number of structurally characterised proteins at least ten-fold. The FSSP database (www2.ebi.ac.uk/dali/fssp/fssp.html), developed by Holm and Sander (1996), provides all-against-all structural comparisons of the proteins with known 3D structures and allows the user to view structural alignments in several convenient formats. This tool is especially useful for identifying and analysing structurally related proteins that have no detectable sequence similarity with each other. The protein structure database at the NCBI (www.ncbi.nlm.nih.gov/ Structure) serves similar objectives.

E. Metabolic Pathways and Classification of Enzymes

The popular scheme of the biochemical pathways, distributed by the Boehringer Mannheim Co., is now available on the WWW at http://expasy.ch/cgi-bin/search-biochem-index. This map can be searched for both the enzyme and metabolite names. It is also linked to the ENZYME database (www.expasy.ch/sprot/enzyme.html), which lists names and catalysed reactions for all the enzymes that have been assigned official EC numbers. A valuable resource for understanding the sets of metabolic reactions in various organisms is provided by the Kyoto Encyclopedia of Genes and Genomes (www.genome.ad.jp/kegg/kegg2.html). This frequently updated site presents a comprehensive set of metabolic pathway charts which conveniently display the lists of enzymes present or apparently absent in each of the completely sequenced genomes. The WIT database (wit.mcs.anl.gov/wit.html/WIT2/), developed by Overbeek *et al.* (1997), is a unique resource for analysis and reconstruction of metabolic pathways from complete or partial genomes. WIT provides functional assignments for nearly 130 prokaryotic and eukaryotic genomes at different stages of completion and is aimed at integrating the metabolic reconstructions within a phylogenetic framework.

F. Taxonomy Database

The taxonomy database (www.ncbi.nlm.nih.gov/Taxonomy/tax.html) at NCBI contains the names of all organisms that are represented in GenBank. It allows the user to browse the universal taxonomic structure and to retrieve protein and DNA sequence data for any particular taxon. The accepted taxonomic structure differs in some respects from the rRNA-based universal tree of life, which can be viewed at www.cme.msu.edu.RDP (Maidak *et al.*, 1997).

G. Integrated Information Retrieval Systems

The problem of providing the user with an easy-to-use interface capable of retrieving various kinds of data from molecular biology data banks is addressed by two sophisticated information retrieval systems, Entrez (Schuler *et al.*, 1996) and Sequence Retrieval System, SRS (Etzold *et al.*, 1996).

Entrez (www.ncbi.nlm.nih.gov/Entrez/) is a search engine that allows users to retrieve molecular biology data and bibliographic citations from the integrated databases maintained at the NCBI. Its most attractive feature is that most of its records are linked to other records, both within a given database and between databases. This allows the user to "jump", for example, from a DNA sequence entry to the corresponding protein entry, check the bibliographic references associated with this sequence, and in some cases, even view the 3D structure of the protein, or the location of the corresponding gene on the chromosome.

Another helpful feature of Entrez is its ability to find documents which are similar to the document the user is looking at. These related documents are called neighbours and can be retrieved by using the "Related Sequences (or Articles)" button. Neighbours for bibliographic references are determined by comparing the title, abstract and indexing terms of each article. Protein and nucleotide neighbours are determined by sequence similarity searches. Since these neighbouring relations are all established at the indexing stage, getting the list of neighbours does not require additional computation and thus occurs very quickly. In the output, the neighbours are listed in the order of relevance, from the closest to the least related ones.

SRS (www.embl-heidelberg.de/srs5/), on the other hand, provides a uniform interface to more than 50 databases at 22 registered sites around the world. The user has to select the databases for further use, and then can follow the links, reaching any data bank from any other data bank through the shortest path. New data banks can be added to the system by creating an appropriate description and indexing the data bank. SRS provides both WWW-based and command-line user interfaces and allows one to conduct complicated queries by applying logical operators for any selected database fields.

◆◆◆◆◆◆ III. SOFTWARE TOOLS FOR SEQUENCE ANALYSIS

Functional prediction for the product(s) of a newly sequenced gene includes identifying the coding sequence, translation of DNA into protein sequence, sensitive similarity searches against various databases, identification of potential motifs and structural features of the protein product, assignment of the probable function and determination, whether this assignment can be considered reliable. Here we shall briefly describe the software tools that are used in each of these stages.

A. From ORFs to Genes

Open reading frames (ORFs) are defined as spans of DNA sequence between start and stop codons. Automatic extraction of all possible ORFs from error-free genomic DNA with known genetic code would seem a straightforward task (which can be performed online at, e.g. www.expasy.ch/www/dna.html or www.ncbi.nlm.nih.gov/gorf/gorf.html). In real life this step is complicated by DNA sequencing errors that may lead to missed or falsely assigned start/stop codons and consequently to extended or shortened ORFs. Given a list of all possible ORFs in a given genome, deciding on which of them constitute genes may be difficult. First, partially or fully overlapping ORFs often occur on the same DNA strand. Second, competing ORFs are commonly present on different DNA strands. Finally, even in the absence of contradictions there is no certainty that an ORF, particularly a short one, actually codes for a protein (Fickett, 1996).

In many cases, genes are identified based on statistically significant sequence similarity of translated ORFs with known protein sequences (Gish and States, 1993). This method is used, for example, in the Analysis and Annotation Tool (http://genome.cs.mtu.edu/aat.html) developed by X. Huang *et al.* (1997). In the absence of significant database hits, gene identification methods based on coding potential assessment and recognition of regulatory DNA elements must be applied. The most widely used program for finding prokaryotic genes, GeneMark (http://genemark.biology.gatech.edu/GeneMark; Borodovsky *et al.*, 1994), employs a non-homogeneous Markov model to classify DNA regions into protein-coding, non-coding, and non-coding but complementary to coding. GeneMark and similar programs (see Fickett, 1996) rely on organism-specific recognition parameters and thus require a sufficiently large training set of known genes from a given organism for successful gene prediction.

Inferring genes by signal and by similarity represent the so-called intrinsic and extrinsic approaches (Borodovsky *et al.*, 1994) , which should ideally be used in combination. The quality of gene prediction can be further improved by using additional available evidence, such as operon structure, location of ribosome-binding sites and predicted signal peptides.

B. Sensitive Similarity Searches

Several algorithms for sequence similarity searches against molecular biology databases have been developed over the years (Altschul *et al.*, 1994; Pearson, 1996). The most widely used of them are Smith–Waterman, FASTA and BLAST, which all offer a reasonable combination of speed and sensitivity. The Smith–Waterman algorithm (Smith and Waterman, 1981) is generally considered the most sensitive of the three; it is also the most time-consuming. Besides, its high search sensitivity often results in increased numbers of false positive hits, which need to be analysed and sorted out by a highly trained biologist. These days, the Smith–Waterman algorithm is often used as a tool of last resort that can detect weak sequence similarities when the other tools fail to do that. The nature and importance of these similarities, of course, have to be critically analysed. The EBI offers similarity searches using the classical Smith–Waterman algorithm (http://croma.ebi.ac.uk/Bic/), or its modified, faster version implemented in the MPsrch program at www.ebi.ac.uk/searches/blitz_input.html.

FASTA (Pearson and Lipman, 1988) is a database search program that achieves sensitivity comparable to that of Smith–Waterman, but is much faster (Pearson, 1996). It is available on the EBI server at www2.ebi.ac.uk/fasta3/. BLAST (Altschul *et al.*, 1990) is the most widely used method of sequence similarity search; it is the fastest one and the only one that relies on a detailed statistical theory (Altschul, 1997). The BLAST suite of programs, available, e.g. at www.ncbi.nlm.nih.gov/BLAST, incorporates three programs that work with nucleotide queries and two programs that use protein queries (Table 1). Actually, only BLASTN performs DNA–DNA comparisons, while the rest compare protein sequences. BLASTN is the most heavily used program of the suite, which is surprising as it is the least sensitive one. In fact, protein sequence comparisons are much more sensitive and should always be preferred to DNA–DNA comparisons. This is especially important for large-scale sequence comparisons that use substantial computer resources (see Baxevanis *et al.*, 1997).

Until recently, the major drawback of the BLAST algorithm has been its slightly lower sensitivity than that of FASTA and Smith–Waterman. In 1996, the BLAST suite of programs was significantly improved by the introduction of new versions that allow gapped alignments, resulting in much higher search sensitivity (Altschul and Gish, 1996). The first version of this set of programs, nicknamed WU-BLAST, is available for database search at www2.ebi.ac.uk/blast2 or www.bork.embl-heidelberg.de, or for downloading from http://blast.wustl.edu.

Recently, Altschul *et al.* (1997) introduced a substantially revised version of the BLAST algorithm that achieves an increased sensitivity at substantially higher search speed. This version of BLAST, referred to as BLAST 2.0, is available as "gapped BLAST" on the NCBI server (www.ncbi.nlm.nih.gov/BLAST) or can be downloaded from ftp://ncbi.nlm.nih.gov/blast/executables. Altschul *et al.* (1997) expect BLAST 2.0 eventually to supersede the previous version as the standard program for most sequence similarity searches.

Finally, in order to achieve even higher search sensitivity, gapped BLAST can be run in an iterative mode, using the PSI-BLAST program (Altschul *et al.*, 1997). This program uses the results of a BLAST output to construct a position-specific scoring matrix, which in turn will be used as a query for the next iteration. This approach achieves the sensitivity of profile-based search methods (Eddy *et al.*, 1995) at substantially lower computation costs. While still under development, this program is

Table 1. Use of BLAST programs for database searches

Program	User-submitted query type	Query type used for database search	Database used for the search
BLASTN	DNA	DNA	DNA
BLASTP	Protein	Protein	Protein
BLASTX	DNA	Translated DNA	Protein
TBLASTN	Protein	Protein	Translated DNA
TBLASTX	DNA	Translated DNA	Translated DNA

already available on the NCBI BLAST server (www.ncbi.nlm.nih.gov/BLAST/).

C. Low Complexity Regions, Non-globular and Coiled-coil Domains

One of the most important advances in database similarity searching during recent years has been the introduction of methods for the automatic masking of low complexity regions. Low complexity regions are basically parts of the protein sequence with local non-random amino acid composition, e.g. rich in glycine or hydrophobic amino acids (Wootton, 1994). In the database similarity search such regions produce multiple hits with other, unrelated, proteins, having similar regions of biased composition. Thus, if a query is a membrane protein, it is likely to produce statistically significant hits with all the membrane proteins in the database. To avoid this and increase the chance of finding true homologues of the given query, such regions should be ignored in the search. The SEG program (Wootton and Federhen, 1996) detects low complexity regions and masks them, substituting X for any amino acids and N for any nucleotides in such a region. The SEG-based filtering is used as default for BLAST searches on the NCBI server; the user has the option of switching it off, though.

The list of protein segments with low compositional complexity also includes non-globular domains, such as myosin rod (Wootton, 1994). The default parameters of the SEG program will not mask non-globular domains; so if the search output contains many hits with non-globular proteins (e.g. myosin), the user should download the program from ftp://ncbi.nlm.nih.gov/pub/seg and run it with adjusted parameters (Wootton, 1994).

Coiled-coil is another protein structural motif that deviates from random distribution of amino acids (Lupas *et al.*, 1991; Lupas, 1996). It represents a bundle of several α-helices arranged in space to form a very stable superhelix. On the sequence level coiled-coil is represented by heptad repeat patterns in which the residues in the first and fourth positions are highly hydrophobic. Prediction of coiled-coil regions of a given sequence can be done with the programs like COILS (Lupas, 1996) at http://ulrec3.unil.ch/software/ COILS_form.html, or Parcoil (Berger *et al.*, 1995) and Multicoil (Wolf *et al.*, 1997) at www.wi.mit.edu/Matsudaira/coilcoil.html.

D. Identification of Sequence Motifs

As noted above, most protein sequence motif databases, such as PROSITE (www.expasy.ch/sprot/prosite.html), BLOCKS (www.blocks.fhcrc.org/), Pfam (www.sanger.ac.uk/Software/Pfam), PRINTS (www.biochem.ucl.ac.uk/bsm/dbbrowser/PRINTS/PRINTS.html), and ProDom (http://protein.toulouse.inra.fr), allow similarity searches against the database. The motifs identified in such searches (e.g. ATP-binding or metal-binding) often allow one to predict the probable function(s) of the unknown protein even in the absence of strong database hits.

E. Structural Characterisation of Protein Sequences

Prediction of structural features for an unknown protein can also be instrumental for identifying its function. Presence of multiple hydrophobic segments often indicates transmembrane topology and can help to attribute the protein to a specific transporter family. In addition, identification of segments with biased composition is necessary for sensitive similarity searches (see above).

Structural analysis of an unknown protein often begins from prediction of a possible signal peptide, which can be done with reasonable accuracy using the SignalP program (Nielsen *et al.*, 1997), available at www.cbs.dtu.dk/services/SignalP.

Once a probable signal peptide has been found, it can be masked, and the sequence can be further analysed for the presence of potential transmembrane segments. Of all the programs that predict transmembrane segments, the statistical method of Persson and Argos (1994) (www.embl-heidelberg.de/tmap/tmap_info.html) and the neural net-based one of Rost *et al.* (1995) (www.embl-heidelberg.de/predictprotein/) appear to perform the best, achieving 95% accuracy in prediction of helical transmembrane segments from multiply aligned sequences.

Prediction of non-globular and coiled-coil domains can also provide an insight into possible functions of an unknown protein (Wootton, 1994; Lupas, 1996).

Sequence-based secondary structure predictions for soluble proteins usually aim at partitioning the protein sequence into one of only three states: α-helix, β-sheet, or a loop. The widely used neural net-based programs PhD (www.embl-heidelberg.de/predictprotein/) by Rost and Sander (1994) and Predator (www.embl-heidelberg.de/argos/predator/predator_info. html) by Frishman and Argos (1997) take advantage of the fact that using a family of related sequences as a query can increase prediction accuracy relative to the single sequence, bringing it to *ca.* 75%. A list of other secondary structure prediction resources is available at http://absalpha.dcrt.nih.gov:8008/otherservers.html

◆◆◆◆◆◆ IV. INTEGRATED SOFTWARE PACKAGES FOR LARGE-SCALE SEQUENCE ANALYSIS

With the development of advanced strategies for genome-scale sequencing, sequence analysis and annotation of complete genomes are becoming the limiting step in most genome projects. It has been estimated that annotating genomic sequence by hand requires as much as one year per megabase (Gaasterland and Sensen, 1996). Hence, considerable efforts are being devoted to automation of the basic steps in this process, i.e. sequence similarity searches and generation of functional predictions for the proteins coded in each particular genome. These projects range in scope from a series of scripts intended to simplify the annotator's work (e.g. SEALS; Walker and Koonin, 1997) to a complete automated system that performs automatic annotation without any human involvement (GeneQuiz; Scharf *et al.*, 1994). Although all these systems are currently used only for in-house projects and are unavailable to outside users, the results produced by these tools are well documented.

A. Comprehensive Genome Annotation Software

The SEALS package (www.ncbi.nlm.nih.gov/Walker/SEALS/) is a modular system of *ca.* fifty convenient UNIX-based tools which follow consistent syntax and semantics. SEALS combines software for retrieving sequence information, scripting database search tools such as BLAST and MoST, viewing and parsing search outputs, searching for protein sequence motifs using regular expressions, and predicting protein structural features and motifs. Using SEALS, the user first looks for the structural features of proteins, such as signal peptides (predicted by SignalP), transmembrane domains (predicted by PHDhtm), coiled-coil domains (predicted by COILS2), and large non-globular domains (predicted using SEG). Once these regions are found and masked, the system looks for regions matching known sequence motifs or matching other known sequences at a high degree of similarity (using BLAST 2.0 and/or PSI-BLAST). Only large globular domains are submitted for BLAST searches, and all the non-identical statistically significant matches are reported for any such search. The final data outputs are intended for use in manual annotation by qualified biologists. SEALS has been extensively used in the comparative studies of bacterial and archaeal genomes (Koonin *et al.*, 1997). Several tools from SEALS are available for downloading from its WWW site.

PEDANT (http://pedant.mips.biochem.mpg.de) is a recently developed WWW resource for exhaustive functional and structural characterisation of proteins encoded in complete genomes (Frishman and Mewes, 1997). For functional assignment of ORFs, PEDANT relies primarily on the results of FASTA and BLAST similarity searches, detection of PROSITE patterns and motifs, and comparisons with conserved sequence blocks. To extract the 3D information, every ORF is compared with the database of secondary structure assignments. Structural classes of globular proteins with unknown 3D structure are suggested on the basis of the secondary structure prediction. Location of membrane-spanning regions, signal peptides, coiled-coil domains and low-complexity segments is delineated using the set of programs listed above. Sequences related to PIR entries are automatically assigned to one of the protein superfamilies and are additionally characterised by PIR keywords. Functional classification of gene products is performed by comparing them with curated master gene sets with assigned functional classes from previously characterised complete genomes. PEDANT makes it possible to create a list of gene products from a given organism belonging to a particular category, e.g. membrane proteins or proteins involved in amino acid metabolism, and then obtain detailed reports on each sequence summarising all known and predicted features. Results of the sequence analysis of proteins from all publicly available complete genomes are available on the PEDANT WWW site.

MAGPIE (www.mcs.anl.gov/home/gaasterl/magpie.html) was designed by Gaasterland and Sensen (1996) as a genome annotation system, accessible by several laboratories working simultaneously on the same project. The system reportedly can change its behaviour and analysis parameters dependent on the particular stage of a sequencing project. MAGPIE assigns confidence levels to multiple features established for each ORF and provides links to associated information, such as bibliographic references and relevant metabolic pathway database entries. MAGPIE is currently being used for annotation of the *Aquifex aeolicus* and *Sulfolobus solfataricus* genomes.

GeneQuiz (www.sander.ebi.ac.uk/genequiz) project represents the first completely automatic system for genome analysis (Scharf *et al.*, 1994) that performs sensitive similarity searches followed by automatic evaluation of results and generation of functional annotation by an expert system based on a set of several predefined rules.

For automated database searches and sequence analysis, GeneQuiz first compares a given ORF against the non-redundant protein database, produced by SRS-assisted linking and cross-referencing of PDB, SWISS-PROT, PIR, PROSITE and TREMBL. This comparison is performed by BLAST and FASTA programs and is used to identify the cases with high similarity, where a possible function can be predicted. Additional searches look for coiled-coil regions, transmembrane segments, and PROSITE patterns (using the programs listed above), perform cluster analysis (Tamames *et al.*, 1997) and secondary structure prediction, and generate multiple alignments. The results are presented as a table that contains information for each ORF on a specified number of best hits (including gene names and database identifiers), predictions for secondary structure, coiled-coils, etc., and a reliability score for each item. The functional assignment is then made on the basis of the functions of the homologues found in the database. At this level, the assignments are qualified as clear, or ambiguous.

The effectiveness of such a system in its current state remains quite uncertain. While Ouzounis *et al.* (1996) estimated the accuracy of their functional assignments to be 95% or better, Koonin *et al.* (1997) reported that only 8 of 21 new functional predictions for *M. genitalium* proteins made by GeneQuiz could be fully corroborated. New functional predictions for the *M. jannaschii* genome reveal similar contrast between the predictions made by the GeneQuiz team (Andrade *et al.*, 1997; see www.sander.ebi.ac.uk/genequiz/genomes/mj/) and those obtained by manual annotation (Galperin and Koonin, 1997; Koonin *et al.*, 1997; see www.ncbi.nlm.nih.gov/Complete Genomes/Mjan2/). Some common pitfalls in functional predictions based on sequence similarities and motifs are listed by Bork and Bairoch (1996), Bork and Koonin (1996) and Galperin and Koonin (1997).

B. Common Features of the Functional Prediction Packages

Although the integrated program suites mentioned above differ in details, the following general framework of large-scale genome analysis software emerges:

- Each system incorporates a locally stored copy of protein and nucleotide databases and a database search engine capable of storing and accessing large amounts of annotated sequence data.
- Functional assignments are primarily based on BLAST and/or FASTA similarity searches against the constantly updated non-redundant protein sequence data bank supplemented by motif searches.
- In many cases certain predicted structural features, such as the number of transmembrane regions or the presence of non-globular domains, serve as the only, albeit weak, indicators of the protein function.
- In contrast to manual, case-by-case analysis where individual decisions on the significance of search hits are possible, automated analysis relies on empirically chosen uniform thresholds that represent a compromise between sensitivity and the number of false assignments.
- Even for highly reliable automatic assignments, analysis by experts remains necessary. Efficient visualisation of results is thus an important prerequisite to successful genome annotation. The most convenient available user interface is an HTML page browser which allows an easy implementation of links between different types of data and is readily suitable for both internal information processing and publicising the results on the Web.

◆◆◆◆◆◆ V. OUTLOOK

The availability of complete genomes adds completely new facets to the sequence analysis work. The new tasks specific for computational genomics include:

- Creating complete functional catalogues of gene products for each particular organism; making definitive conclusions about the presence or absence of certain proteins (hence, functions, metabolic pathways, etc.).
- Examining general organisation of the gene complement (e.g. gene order, operon architecture); assessing the redundancy of genetic information.
- Conducting cross-genome comparisons to delineate characteristic features of particular organisms and/or taxons (e.g. identify virulence-related proteins in pathogens).

New experimental data or functional assignments produced in the course of one project can often be used to improve the results of another project, making genome annotation a never-ending iterative process. The task of continuously updating the information pertinent to a particular organism or a group of related organisms will likely be taken on by specialised databases, maintained by and serving the needs of scientists studying these organisms. Such databases, which have already been created for *Escherichia coli* (http://ecocyc.PangeaSystems.com/ecocyc/ecocyc.html

and http://mol.genes.nig.ac.jp/ecoli), *Saccharomyces cerevisiae* (www.mips.biochem.mpg.de/mips/yeast/ and genome-www.stanford.edu/Saccharomyces) and *Bacillus subtilis* (www.pasteur.fr/Bio/SubtiList.html and http://ddbjs4h.genes.nig.ac.jp), complement the sequence data with the biochemical, genetic and ecological information extracted from the literature. Similar comprehensive databases are being planned for other organisms with completely sequenced genomes, which should benefit both academic studies and medical research.

Another important direction of post-genomic analysis is the reconstruction of metabolic pathways present or absent in a particular organism. The WIT database (wit.mcs.anl.gov/wit.html/WIT2/) allows one to search any of the completed genomes for the likely candidates that could take on the functions of the missing enzymes. The enzyme functions for which no candidates can be found would indicate cases of non-orthologous gene displacement (Koonin *et al.*, 1996a) or identify missing pathways.

Finally, comparisons of complete genomes allow one to identify definitively orthologous genes and proteins in different phylogenetic lineages, which not only helps to understand biochemical evolution, but also indicates the likely function(s) of all the members of each such cluster of orthologous genes (Tatusov *et al.*, 1997; see www.ncbi.nlm.nih.gov/ COG).

The history of sequence annotations of complete genomes shows that even when the whole arsenal of the available tools is used to gain as much functional information as is currently possible, a substantial fraction of gene products, from 25 to 30% (Koonin *et al.*, 1997) up to 60% (Bult *et al.*, 1996), remains totally uncharacterised. Uncovering the functions of these remaining proteins, as well as identifying the precise roles of others for which only a general prediction could have been made, will be possible only by direct experimental approaches, e.g. by disrupting respective genes and analysing resulting mutant phenotypes. Such projects for *E. coli* and yeast are already under way and promise eventually to bring us to the next milestone in genome analysis – a complete functional description of all the genes in an organism.

Acknowledgements

The opinions expressed in this chapter do not necessarily reflect the positions of the NCBI or MIPS. We thank Eugene Koonin, Renata McCarthy and Francis Ouellette (NCBI) and Hans-Werner Mewes (MIPS) for critical reading of the manuscript.

References

Altschul, S.F. (1997). Sequence comparison and alignment. In *DNA and Protein Sequence Analysis: A Practical Approach* (M. J. Bishop and C. J. Rawlings, eds), pp. 137–167. IRL Press, Oxford.

Altschul, S. F. and Gish, W. (1996). Local alignment statistics. *Meth. Enzymol.* **266**, 460–480.

Altschul, S. F., Gish, W., Miller, W., Myers, E. W. and Lipman, D. J. (1990). Basic local alignment search tool. *J. Mol. Biol.* **215**, 403–410.
Altschul, S. F., Boguski, M. S., Gish, W. and Wootton, J. C. (1994). Issues in searching molecular sequence databases. *Nature Genet.* **6**, 119–129.
Altschul, S. F., Madden, T. L., Schaffer, A. A., Zhang, J., Zheng, Z., Miller, W. and Lipman, D. J. (1997). Gapped BLAST and PSI-BLAST – a new generation of protein database search programs. *Nucl. Acids Res.* **25**, 3389–3402.
Andrade, M., Casari, G., de Daruvar, A., Sander, C., Schneider, R., Tamames, J., Valencia, A. and Ouzounis, C. (1997). Sequence analysis of the *Methanococcus jannaschii* genome and the prediction of protein function. *Comput. Appl. Biosci.* **13**, 481–483.
Attwood, T. K., Beck, M. E., Bleasby, A. J., Degtyarenko, K., Michie, A. D. and Parry-Smith, D. J. (1997). Novel developments with the PRINTS protein fingerprint database. *Nucl. Acids Res.* **25**, 212–217.
Bairoch, A. and Apweiler, R. (1997). The SWISS-PROT protein sequence data bank and its supplement TrEMBL. *Nucl. Acids Res.* **25**, 31–36.
Bairoch, A., Bucher, P. and Hofmann, K. (1997). The PROSITE database, its status in 1997. *Nucl. Acids Res.* **25**, 217–221.
Baxevanis, A. L., Boguski, M. S. and Ouellette, B. F. F. (1997). Computational analysis and annotation of sequence data. In *Genome Analysis: A Laboratory Manual* (B. Birren, E. D. Green, S. Klapholz, R. M. Myers and J. Roskams, eds), vol. 1, pp. 533–586. CSHL Press, Cold Spring Harbor.
Berger, B., Wilson, D. B., Wolf, E., Tonchev, T., Milla, M. and Kim, P. S. (1995). Predicting coiled coils by use of pairwise residue correlations. *Proc. Natl. Acad. Sci. USA* **92**, 8259–8263.
Bork, P. and Bairoch, A. (1996). Go hunting in sequence databases but watch out for the traps. *Trends Genet.* **12**, 425–427.
Bork, P. and Gibson, T. J. (1996). Applying motif and profile searches. *Meth. Enzymol.* **266**, 162–184.
Bork, P. and Koonin, E. V. (1996). Protein sequence motifs. *Curr. Opin. Struct. Biol.* **6**, 366–376.
Borodovsky, M., Rudd, K. E. and Koonin, E. V. (1994). Intrinsic and extrinsic approaches for detecting genes in a bacterial genome. *Nucl. Acids Res.* **22**, 4756–4767.
Bult, C. J., White, O., Olsen, G. J., Zhou, L., Fleischmann, R. D., Sutton, G. G., Blake, J. A., FitzGerald, L. M., Clayton, R. A., Gocayne, J. D. *et al.* (1996). Complete genome sequence of the methanogenic archaeon, *Methanococcus jannaschii. Science* **273**, 1058–1073.
Eddy, S. R., Mitchison, G. and Durbin, R. (1995). Maximum discrimination hidden Markov models of sequence consensus. *J. Comput. Biol.* **2**, 9–23.
Etzold, T., Ulyanov, A. and Argos, P. (1996). SRS: information retrieval system for molecular biology data banks. *Meth. Enzymol.* **266**, 114–128.
Fickett, J. W. (1996). Finding genes by computer: the state of the art. *Trends Genet.* **12**, 316–320.
Fleischmann, R. D., Adams, M. D., White, O., Clayton, R. A., Kirkness, E. F., Kerlavage, A. R., Bult, C. J., Tomb, J. F., Dougherty, B. A., Merrick, J. M. *et al.* (1995). Whole-genome random sequencing and assembly of *Haemophilus influenzae* Rd. *Science* **269**, 496–512.
Fraser, C. M., Gocayne, J. D., White, O., Adams, M. D., Clayton, R. A., Fleischmann, R. D., Bult, C. J., Kerlavage, A. R., Sutton, G., Kelley, J. M. *et al.* (1995). The minimal gene complement of *Mycoplasma genitalium. Science* **270**, 397–403.
Frishman, D. and Argos, P. (1997). 75% accuracy in protein secondary structure prediction. *Proteins* **27**, 329–335.

Frishman, D. and Mewes, H. W. (1997) PEDANTic genome analysis. *Trends Genet.* **13**, 415–416.

Gaasterland, T. and Sensen, C. W. (1996). Fully automated genome analysis that reflects user needs and preferences. A detailed introduction to the MAGPIE system architecture. *Biochimie* **78**, 302–310.

Galperin, M. Y. and Koonin, E. V. (1998). Hurdles on the road to functional annotation of genomes: domain rearrangement, non-orthologous gene displacement, and operon disruption. *In Silico Biology* **1** (in press).

George, D. G., Dodson, R. J., Garavelli, J. S., Haft, D. H., Hunt, L. T., Marzec, C. R., Orcutt, B. C., Sidman, K. E., Srinivasarao, G. Y., Yeh, L. S. L. *et al.* (1997). The Protein Information Resource (PIR) and the PIR-International Protein Sequence Database. *Nucl. Acids Res.* **25**, 24–28.

Gish, W. and States, D. J. (1993). Identification of protein coding regions by database similarity search. *Nature Genet.* **3**, 266–272.

Goffeau, A., Barrell, B. G., Bussey, H., Davis, R. W., Dujon, B., Feldmann, H., Galibert, F., Hoheisel, J. D., Jacq, C., Johnston, M. *et al.* (1996). Life with 6000 genes. *Science*, **274**, 546, 563–567.

Gouzy, J., Corpet, F. and Kahn, D. (1996). Graphical interface for ProDom domain families. *Trends Biochem. Sci.* **21**, 493.

Henikoff, J. G., Pietrokovski, S. and Henikoff, S. (1997). Recent enhancements to the Blocks database servers. *Nucl. Acids Res.* **25**, 222–225.

Holm, L. and Sander, C. (1996). The FSSP database: fold classification based on structure–structure alignment of proteins. *Nucl. Acids Res.* **24**, 206–209.

Huang, X., Adams, M. D., Zhou, H. and Kerlavage, A.R. (1997). A tool for analyzing and annotating genomic sequences. *Genomics* **46**, 37–45.

Koonin, E. V., Mushegian, A. R. and Bork, P. (1996a) Non-orthologous gene displacement. *Trends Genet.* **12**, 334–336.

Koonin, E. V., Tatusov, R. L. and Rudd, K. E. (1996b). Protein sequence comparison at genome scale. *Meth. Enzymol.* **266**, 295–322.

Koonin, E. V., Mushegian, A. R., Galperin, M. Y. and Walker, D. R. (1997). Comparison of archaeal and bacterial genomes: computer analysis of protein sequences predicts novel functions and suggests a chimeric origin for the archaea. *Mol. Microbiol.* **25**, 619–637.

Lupas, A. (1996). Prediction and analysis of coiled-coil structures. *Meth. Enzymol.* **266**, 513–525.

Lupas, A., Van Dyke, M. and Stock, J. (1991). Predicting coiled coils from protein sequences. *Science* **252**, 1162–1164.

Maidak, B. L., Olsen, G. J., Larsen, N., Overbeek, R., McCaughey, M. J. and Woese, C. R. (1997). The RDP (Ribosomal Database Project). *Nucl. Acids Res.* **25**, 109–111.

Mewes, H. W., Albermann, K., Bahr, M., Frishman, D., Gleissner, A., Hani, J., Heumann, K., Kleine, K., Maierl, A., Oliver, S. G. *et al.* (1997a). Overview of the yeast genome. *Nature* **387**, 7–65.

Mewes, H.W., Albermann, K., Heumann, K., Liebl, S. and Pfeiffer, F. (1997b). MIPS: a database for protein sequences, homology data and yeast genome information. *Nucl. Acids Res.* **25**, 28–30.

Murzin, A. G., Brenner, S. E., Hubbard, T. and Chothia, C. (1995). SCOP: a structural classification of proteins database for the investigation of sequences and structures. *J. Mol. Biol.* **247**, 536–540.

Nielsen, H., Engelbrecht, J., Brunak, S. and von Heijne, G. (1997). Identification of prokaryotic and eukaryotic signal peptides and prediction of their cleavage sites. *Protein Eng.* **10**, 1–6.

Ouellette, B. F. F. and Boguski, M. S. (1997) Database divisions and homology search files: a guide for the perplexed. *Genome Res.* **7**, 952–955.

Ouzounis, C., Casari, G., Valencia, A. and Sander, C. (1996). Novelties from the complete genome of *Mycoplasma genitalium*. *Mol. Microbiol.* **20**, 898–900.

Overbeek, R., Larsen, N., Smith, W., Maltsev, N. and Selkov, E. (1997). Representation of function: the next step. *Gene* **191**, GC1–GC9.

Pearson, W. R. (1996). Effective protein sequence comparison. *Meth. Enzymol.* **266**, 227–258.

Pearson, W. R. and Lipman, D. J. (1988). Improved tools for biological sequence comparison. *Proc. Natl. Acad. Sci. USA* **85**, 2444–2448.

Persson, B. and Argos, P. (1994). Prediction of transmembrane segments in proteins utilising multiple sequence alignments. *J. Mol. Biol.* **237**, 182–192.

Peruski, L. F. and Peruski, A. H. (1997). The Internet and the new biology: tools for genomic and molecular research. ASM, Washington, DC.

Rost, B. and Sander, C. (1994). Combining evolutionary information and neural networks to predict protein secondary structure. *Proteins* **19**, 55–72.

Rost, B., Casadio, R., Fariselli, P. and Sander, C. (1995). Transmembrane helices predicted at 95% accuracy. *Protein Sci.* **4**, 521–533.

Sander, C. and Schneider, R. (1991). Database of homology-derived protein structures and the structural meaning of sequence alignment. *Proteins* **9**, 56–68.

Scharf, M., Schneider, R., Casari, G., Bork, P., Valencia, A., Ouzounis, C. and Sander, C. (1994). GeneQuiz: a workbench for sequence analysis. *Intell. Syst. Mol. Biol.* **2**, 348–353.

Schuler, G. D., Epstein, J. A., Ohkawa, H. and Kans, J.A. (1996). Entrez: molecular biology database and retrieval system. *Meth. Enzymol.* **266**, 141–162.

Smith, T. F. and Waterman, M. S. (1981). Identification of common molecular subsequences. *J. Mol. Biol.* **147**, 195–197.

Sonnhammer, E. L. L. and Kahn, D. (1994). The modular arrangement of proteins as inferred from analysis of homology. *Protein Sci.* **3**, 482–492.

Sonnhammer, E. L. L., Eddy, S. R. and Durbin, R. (1997). Pfam: a comprehensive database of protein domain families based on seed alignments. *Proteins* **28**, 405–420.

Tamames, J., Casari, G., Ouzounis, C. and Valencia, A. (1997). Conserved clusters of functionally related genes in two bacterial genomes. *J. Mol. Evol.* **44**, 66–73.

Tatusov, R. L., Koonin, E. V. and Lipman, D.J. (1997). A genomic perspective on protein families. *Science* **278**, 631–637.

Walker, D. R. and Koonin, E. V. (1997). SEALS: a system for easy analysis of lots of sequences. *Intell. Syst. Mol. Biol.* **5**, 333–339.

Woese, C. R. (1994). There must be a prokaryote somewhere: microbiology's search for itself. *Microbiol. Rev.* **58**, 1–9.

Wolf, E., Kim, P. S. and Berger, B. (1997). MultiCoil: a program for predicting two- and three-stranded coiled coils. *Protein Sci.* **6**, 1179–1189.

Wootton, J. C. (1994). Non-globular domains in protein sequences: automated segmentation using complexity measures. *Comput. Chem.* **18**, 269–285.

Wootton, J. C. and Federhen, S. (1996). Analysis of compositionally biased regions in sequence databases. *Meth. Enzymol.* **266**, 554–571.

Index

Page numbers in *italics* refer to figures and tables

Index

Index